Catalogue of selected seashells from the Emilio F. García collection. And other stories.

Emilio F. García

With more than 500 images, mostly from the
Gulf of Mexico and the Caribbean

ISBN: 979-889766948-6

TABLE OF CONTENTS

INTRODUCTION

Why do I still remember the first shell I ever picked up speaks for the attractive power seashells had on me from a very young age. I was about 4 years old and my parents had taken me to the Bay of Pigs, only a few miles from Jagüey Grande, my hometown. However, to get to the bay we had to cross a section of the Zapata Swamp in a bus-cum-train-wheels that had to move very slowly. The narrow gauge rails were in large part submerged under several inches of water; a slow but fascinating trip for one who loves nature. And at the end of the journey we still had to walk (run, really) for a quarter of a mile to get to the beach; too many mosquitos to walk idly. Oh, yes, that first shell was a single valve of *Tellina radiata.*

I took numerous trips to the Bay of Pigs in later years, always looking forward to collecting shells. Then, when I was eight, an uncle who knew very well my passion, took me to Havana to see the collection of the most famous Cuban malacologist, Don Carlos de la Torre y Huerta. Don Carlos had already died but one of his relatives, friend of my uncle, showed me the collection, and also gave me a booklet on Cuban shells identified by their "scientific" names. I was thrilled to be able to give names to my shells. And the rest is history.

Well, not quite, because at twelve years of age I was sent to a boarding school and, upon graduation, to the United Sates. No time for collecting, at least not until 1965, when I started teaching at the University of Southwestern Louisiana (now University of Louisiana at Lafayette). That gave me some time to start collecting again (since I had lost my Cuban collection), first in Louisiana, later in Florida, and eventually around the world. At the very beginning I thought I was the only collector. In a way, as was lucky, because every new shell I found was a real treasure. I don't know if I could have handled it if my first experience had been a shell bourse at a convention.

So, from the rather distant past to the present, I started this project as a personal assignment, and had it not been for a colleague at the university who suggested that it be made more public, the project would have stayed in my computer. The reader must forgive me if my name appears way too many times in this work. It is in an unorthodox format, whimsical, as the original work was meant to be private, sort of like a stream of consciousness style in literature. And once I decided on that track, I had no inclination to do a more formal presentation. However, even in this format the work should be of some use.

The plates represent mostly Gulf of Mexico and Caribbean species, as these two regions have been my area of concentration. They are especially focused below. Now and then you will be surprised by a "foreign" species; just a whim, I was entertaining myself.

The captions below the 55 plates include the name of the species (according to me), size, and respective catalogue number in parenthesis. The catalogue numbers will appear in the back of the work, in numerical sequence, as "Index", with substantial locality data. I have omitted "State", "Province" and "Date" of collection, as the lines would be too long for this format. I have also provided some pertinent (and impertinent) comments on a number of the figures. These comments are, again, in numerical order under "Comments on some figures". And, finally, there are the "References", all of which have to do with me. Sorry. For those who are really curious I have attached an "Appendix" with my Curriculum Vitae at the end of the catalogue.

I have two special sections for Roatán Island, northern Honduras, and for the Gulf of Mexico campaign on board the R/V *Pelican*. Both areas have been of special interest to me for several decades. My collection has 1032 lots obtained in Roatán, although some specimens may have been collected by the fishermen in Nicaragua, and even Colombia. The Gulf of Mexico section includes specimens bequeathed to me by Frank Frumar, of Kirkwood, Missouri, an avid collector who dredged for many years in deep water off the Florida Keys; the collection of Thomas and Barbara McGinn, whose collection I obtained, and after whom *Scaphella mcginnorum* was named; and the results of the many campaigns on board the R/V *Pelican*. All together, the collection has over 4,600 lots from these three sources, which is about 15.3 % of the entire collection.

As the title implies, only selected specimens are shown. They are arranged in taxonomic order but with an unorthodox reasoning for the choice of species. In some cases, a number of specimens are represented to show what I consider variability of the same species. Let me warn you, I tend to be a "lumper".

ROATAN ISLAND AND THE WESTERN CARIBBEAN

Mostly everyone in the Biology Department at my university knew that I collected shells, and one day a graduate student showed up in my office to tell me that he was going SCUBA diving to an island off the northern coast of Honduras called Roatán; could he bring me some shells? I never refuse such invitations but I also did not want him to burden himself with useless, common material so I showed him a picture of a *Conus granulatus* and said: that's what I want. And he brought it to me; and he also brought a number of other rare shells that I did not know were there. So I said to myself, I have to go.

And so a couple of years passed when this beautiful brunette girl with green eyes, Annie Jones, showed up at my office; she wanted to major in Spanish. She spoke with an accent but the accent was Caribbean, not of a Spanish speaker, and when I asked her where she was from she said Honduras; and my antennas came up. I told her I had been wishing to go to Honduras for a while, but to an island where she had probably never been, called Roatán. I know it, she said. I am from there. Are there any shrimpers or lobster fishermen in the island? I asked. Sure, she said, my father, my uncles, my brother… And so I explained my interest in seashells and why I asked those questions. Well, she said, I am going to visit my parents at the end of the semester, would you like to come? And, no, she was not my student.

So next May Annie, her friend Michelle, my friend Mike Lee and I flew from Lafayette to Houston to La Ceiba, in the northern coast of Honduras. There, with luck, we would get on a DC-3 airplane and make the short flight to Roatán. We were lucky, it happened. The building airport was a dilapidated wooden structure where the luggage was checked (not really) in a 12 ft by 12 ft. room with a counter. You should see the place today!

Transport around Roatán at that time was almost non-9existent, and the dirt roads were nominally usable, almost impossible to navigate during the rainy season. But no worries, because Babs, Annie's mother, was that night waiting for her daughter with a minibus that she had hired; a cab would not do; Annie was expected to bring (and did) many things for her very large extended family.

Our destination was Jonesville, a small village in south-central Roatán named after Annie's family. Queen Victoria had officially signed a document granting that portion of the island to the Joneses. George Jones, Annie's father, showed me the parchment. Not only was George the captain of a shrimp boat but, as Annie had said, his brothers, son and most of Jonesville were shrimpers. And, of course, they knew shrimpers and lobster fishermen from everywhere on the island. Babs had a banquet for her daughter, whom she had not seen for a year. She was a great cook, and among her culinary delights were such exotic dishes as "malanga" pie and iguana tails. Lobster, of course, was served routinely.

That first trip to Roatán was an incredible success, both for the variety and quality of my experiences, as for the number of rare shells that I obtained, some of which were unknown to science. Mike and I would return one more time to Jonesville, where we would stay at Annie's grandmother's empty house. The village was about two blocks long, and "our" house was at the opposite end from Annie's parents, so we walked there for breakfast and dinner. Now everyone knew us by name and stopped us for visits. Some evenings we played volleyball in front of the one little grocery store where a net was set up. It was idyllic. Not only did we get great shells, we had gained a village.

Considering that we might be taking advantage of the Joneses hospitality, in future trips we started lodging in a "motel" (to be generous) in Oak Ridge, a larger town next to Jonesville. The disadvantage was that the little 3-room "motel" was a hole in the wall with clouds of mosquitos. The advantages were that we could take the cheap island busses to go almost anywhere in the island for one dollar; also, there was a fleet of fishermen there (see fig. C). The motel was owned by an island lady named Goldie Cooper, who was Annie Jones' aunt (almost everyone in that area was related, and they called each other "primo", cousin), so we were welcome at her house, a few steps from our room, for morning coffee and nightly libations. She had her own little "dory", so many nights we would go across the canal to the town bar called "The Happy Landing". I remember one night there was a guest who was the guide for a group of Americans who were visiting the island. He wanted to get away from them for a while so he took the night off to join us. He had brought his own tiny boat to Goldie's, and we had taken her larger boat to "The Happy Landing". Upon our return, the fellow's boat was no longer there. It had not been tied properly and had drifted away. The bad news was that all the passports from the people in his group were on the boat. Eventually, in the middle of the night, the boat and passports were found next to someone else's pier.

I travelled to Roatán two or three times a year until I retired in 1997. The mosquitos had won the battle at Goldie's place, but I had discovered another small (6-room) "hotel" in a key (fig. B) just off Oak Ridge and stayed there many, many times. A few minutes ride in a dory (fig. A) would take me to Hotel San José, managed by Tina Reyes, a very intelligent and friendly lady from the mainland. We became very good friends, and I would bring her from the States items that she needed or that I thought she might want: a blender here, a little camera there, etc. In return she would cook for me the most fabulous meals. You see, she had two brothers who were fishermen, and would bring her shrimp, lobster, stone crab, conch, etc. Of course, they would also bring great shells that I would buy from them. My trips there were so frequent that I finally decided to leave a suitcase at the main house with essentials, such as my diving equipment, a reading light that I hooked behind my bed, sandals, collecting material, etc. By that time I had acquired a network of shrimpers, lobster fishermen and conch fishermen who kept shells for me. One of the conch fishermen, who had a fleet of boats, had a collection of conch pearls of some two hundred specimens of many colors worthy a museum.

The hotel San José itself could not have been more basic. There were three rooms on either side of the old wooden building, with a narrow hall separating the two sides. The last corner rooms had a private shower (no hot water) and a fan; the other two rooms on either side had a sharing bathroom and no fans. I did not mind at all the simplicity of the hotel; the only drawback was that the intervening walls between rooms did not go all the way up to the ceiling so there was a certain lack of privacy if there were other guests. Fortunately, most of the time I was there by myself. My morning coffee, breakfast and dinner were at the main house. I never had lunch because, once I left the key after having breakfast, I would disappear until late afternoon. I remember one of those late afternoons, when I had come with a smelly *Charonia* and went across the path to the beach to clean it. As I was shaking the shell a *Morum oniscus* decided to come from its hiding place and investigate. Although I had collected a number of empty specimens, I had never seen a live one.

I eventually became a good friend of Don José, the owner. He was a gentle old man, and most mornings, when I went for coffee and breakfast to the main house, we would sit in his back porch; he smoking his cigarettes and I drinking my coffee, and he would tell me stories from “the old times”, when there were few roads in the island and one had to walk everywhere. His stories from “back when” were delightful.

After breakfast I had a full day of adventure. I would take one of the many little dories that passed through the canal that separated the key from the island (fig. B), and then take one of the busses that went through different towns; or just start hiking. There was a quaint small village on the north side called Punta Gorda, and at the time there were very many fishermen that lived there. It was only a couple of miles between Oak Ridge and Punta Gorda, so, if it wasn’t raining, I would just start walking, enjoying the scenery, the flying parrots, the silence. “Never more”, said the raven; now it is all houses built by Hondurans from the mainland looking for a better way of life.

I discovered the richness of Punta Gorda by chance. It was Babs’ birthday, and we wanted to reciprocate her gracious hospitality by taking the whole family (yes, including Goldie) to a nice resort called Spy Glass Hill. Happy, the owner, had a great display of shells in her dining room and we started talking shells. That was my first sight of *Pleioptygma helenae*, the endemic species that inhabits a small area in the western Caribbean. This species, of which she had several specimens, and other great shells in her collection, had come from Punta Gorda, just down the hill from the resort. From then on I went to Punta Gorda so many times that eventually I did not have to find people with shells, they would find me; I was “the shell man”. I became well acquainted with a small boy, Joseph, who kept following me in one of the trips, and he would tell me who had come from the fishing banks and who might have shells; and he would take me to their house. So, whenever I arrived in Punta Gorda from a new trip, I would go to his house, and he would take me to the houses of the recently arrived fishermen.

On one occasion, on my third visit to the village, little Joseph came running to me very excited. He said that someone had arrived from a extensive trip to Colombia and had many shells, but he did not live in Punta Gorda itself. It took us a long time to get there; the family had built a small thatch-roofed house on a hill in the middle of nowhere. When we finally arrived, the dogs started barking so the mother of the fisherman come out of the hut with a suspicious look on her face at the unexpected visitor. Yes, her son had just come from the fishing banks in Colombia; and, yes, he had many shells but he was sleeping. Pressing my point that perhaps her son could make good money from the shells he brought, she woke him up, but he was not happy. Moreover, he had already promised the shells to someone in French Harbor, a small fishing town in the south coast. I asked him if I could see the shells, and he obliged, and when I saw the shells I knew I had him. I will never forget the look on his face when I told him how much I could offer for some of them (e.g., fig. 49). I did not buy all the shells but, on second thought, the next day I came back and got the rest. We were all happy.

French Harbor is the center for the shrimp and lobster fishing in Roatán Island, Oak Ridge being the second (fig. C), and the fishermen brought back many shells. When I first started going to the island I published an article about the molluscan riches of the area, and many collectors began visiting the island. There were a couple of native entrepreneurs who saw what was going on and bought shells from the fishermen for very little and sold them to tourists and people like me so, of course, I also went through French Harbor buying shells from them and other fishermen who were my personal contacts. There, as in Punta Gorda, they knew me as "the shell man", a couple of them even called me by my name.

I went alone for most of my last trips before 1997. Mike knew the island very well by then, and buying shells from the fishermen was not his cup of tea. Besides, in the rainy season it was very difficult to leave the island because the road from Oak Ridge to the airport was impassable. During my many trips there I had become a good friend with the owner and the manager of a beautiful resort at the west end of the island called Anthony's Key. During those times, when it was hard to make it to the airport from Oak Ridge, I would spend the last night at this fantastic resort for free; and to reciprocate, I would present the guests with a slide show of the seashells of the Bay Islands. I even set up a nice display of seashells in their museum (now gone). I was during one of my stays there when I experienced my first earthquake. Since the bungalows are on stilts, the experience was more dramatic than it was in reality so, other than the few seconds of shake, nothing happened.

When I decided to go back to Roatán a few years after my retirement I contacted Annie, my island friend, and the news was sad. Don José had died; his heirs had not taken proper care of the building, and it had rotted away; and Tina Reyes had gone back to her home in the mainland. So, my island friends found for me a beautiful home just across the old San José, and it was wonderful: a two-story building with a large balcony overlooking the ocean. On clear days, seeping a some wonderful drink on the balcony, one could see across the waters the mountains on the mainland, and later on the magnificent sunsets to the west.

The housekeepers, Victor and Marta, lived downstairs. Marta cooked for us; Victor took care of whatever plans we had for the next day. He used to own a dory, and in some of my earlier trips he used to take me snorkeling to my favorite spot (see fig. D), which was in itself an adventure. Another plus was that Victor knew most fishermen in Oak Ridge; I would tell Victor when I was coming, and he in turn would tell the fishermen, and they would come to the house with their treasures (see figs. E-F). Also, since I did a lot of walking through the villages, I would tell other fishermen, or their families, where I was staying, and when I would be back to the island.

Yes, Roatán at that period was a paradise for a shell collector, but those days are gone. The island became a tourist spot -there were not even telephones in my early days- and with the tourists came the big ships. So, why should someone be a hard-working, poorly paid deck hand when he could be driving one of the very many taxis, or be a waiter at any of the multiple resorts that sprung up throughout the island. Also, with all the new construction came the silting of the water and the ruination of habitats. In one of my last trips to Roatán I visited my favorite spots where Victor used to take me in his dory. I only found a small *Caribachlamys pellucens (=imbricata).* And the rest **is really** history.

ROATAN ISLAND- A- Catching a dory to go to "the key"., where our house was located. B- View of "the key" from a near-by hill. C- Fleet of shrimp/lobster boats. D- Mangrove canal, supposedly made by buccaneers to be able to flee from the British navy. We would take this canal to go to my favorite collecting place. E_F- Fishermen bringing their ware to the house.

THE R/V *PELICAN* AND THE GULF OF MEXICO

The R/V *Pelican* (fig. H) is the flagship of the Louisiana Universities Marine Consortium (LUMCON), located in Cocodrie, south Louisiana. It is 116 ft. long and is designed for oceanographic research missions, including dredging; it has a dredge winch that can hold from 8,000 to 10, 000 meters of cable, depending on the thickness.

In 1993 Dr. Darryl Felder, then head of the Biology Department at the University of Louisiana at Lafayette invited me to join one of his campaigns on board the *Pelican*; he was doing extensive work on crustaceans of the Gulf of Mexico (GoMx) but wanted someone to take care of the mollusks. The invitations continued until 2019, first by Dr. Felder and later by Dr. Suzanne Fredericq, recognized worldwide for her work on red algae. I also thank Dr. Fredericq for helping me with technical issues during the compilation of this catalogue.

Life on the *Pelican* was very exciting. After bringing on board all necessaries, we had lectures about safety, and then down (or up to our bunks to go to sleep because the cruise would not start until 12:01 a.m. It would take all night to reach our destination, usually from 6 a.m. to 8 a.m. depending on where we were going. Meals were served at 6 a.m., 12 p.m. and 6 p.m. I usually skipped breakfast since I only needed cereal and milk, which were always available; as were large refrigerated compartments with many other goodies, such as ice cream bars, etc., etc. We also had the good fortune on many cruises to have a cook of *cordon bleu* quality and had incredibly sumptuous meals; sometimes fish caught by the crew or shrimp caught in the dredge.

The campaigns mainly targeted crustaceans, algae and mollusks, but other interesting marine life, such as fish or echinoderms, was kept in the freezers or in ethanol to send to workers in other fields. When the dredges came up the group, composed of faculty and graduate students alike, could not wait to inspect the haul, but we did have to wait, ready with life jackets and head gear, ready with our collecting trays, until the crew considered it safe for us to run out to the deck and get around the table where the material was deposited (figs. I, J, L). So, crustaceans went on to some buckets and trays, algae to other buckets and trays and mollusks to mine, since I was the only individual collecting mollusks.

And it was a sight to be held because if any of us found a specimen of species other than that of our own interest, we would pass it to the pertinent depositories. And the OHs! and AHs! were heard all over the place when rare or unknown specimens, totally insignificant, let's say, for someone interested in crustaceans, turned into a resplendent diamond in the hands of the algae or mollusk "team".

And at times "miracles" did happen. There was a crew member who was a magician with machines and gadgets but could care less about what we were doing and kept always very much to himself. He fixed many a dredge that was damaged during the expeditions. However, eventually he was attracted to our sorting table where all the excitement was happening and started to pick up a shell here and a shell there; so, after he handed me a couple of interesting specimens and saw my reaction, he was hooked. The most memorable moment was when I was reading in the galley, unaware that another dredge had come up, and here he comes with a big smile on his face and a *Haliotis pourtalesii*, a rare species, in his hand. We even embraced!

But all that happened when we were dredging with the box dredge in the relatively shallow banks (50-100 m). It was a different story when the *Pelican* went to deep water and used the Benthic Skimmer. It was a long way to the sites. It was a long hour to set the dredge down, then one hour of dredging and then another long hour up. So, sometimes it was 2 a.m. or 3 a.m. when the Benthic Skimmer finally got on deck. But, of course, it was all worth it.

In the 23 years that we dredged on the *Pelican,* most of the northern and northeastern GoMx were sampled, as well as Bahía de Campeche, northern Yucatán Peninsula, Mexico (see map, fig. G). For the shallower bottoms (less than 100 m), usually composed of coralline algal hard banks occurring off Louisiana and elsewhere, we used the box dredge (see figs. I-J). For the deeper mud and silt bottoms, a specially designed dredge for deep water, the Benthic Skimmer, was used (see figs. K-L; also García, 2007b). Reports on the findings of the dredging expeditions have been reported elsewhere (see García, 1999a, 1999b, 2000, 2002a, 2002b, 2005, 2006a, 2006b, 2007a, 2007b, 2007c, 2007d, 2008a. 2008b, 2008c 2008d, 2008e, 2008f, 2009, 2011a, 2011b, 2011c, 2012, 2013a, 2018, 2019, 2021a, 2021b, 2022a, 2023 a, 2023b, 2023c, 2023e, 2023f, 2023g, 2024b, 2024b, 2024d; and García and Lee, 2002, 2003, 2004).

From these campaigns, 4154 lots have been catalogued: 1952 from Louisiana, 57 from Mississippi, 430 from Alabama, xxx from Texas, 810 from Florida, and 905 from Bahía de Campeche, Mexico. Most campaigns occurred off the Louisiana coast, thus the high number of lots from that state. However, the single expedition to Campeche, which lasted two weeks, produced the highest proportional number of lots. Although some hauls were made as shallow as 20 m, most were made in deeper water, from approximately 60m to 1800m. It would take the *Pelican* most of a day to get to the deeper stations, lower the dredge, do the dredging, and bring back up the Skimmer. On several occasions we lost the box dredge, and on one occasion the Benthic Skimmer; a very frustrating affair.

Among the specimens dredged by the *Pelican* there were 29 species new to science:

Conus sauros García, 2006

Cylindriscala rosenbergi García, 2005

Neoterebra mugridgeae (García, 1999)

Opaliopsis rabalaisi García, 2005

Gymnobela felderi García, 2005
Gymnobela fredericqae García, 2005
Gymnobela petiti García, 2005
Anticlinura atlantica García, 2005
Turritella lyonsi García, 2006
Monostiolum harryleei García, 2006
Cosmioconcha geigeri García, 2006
Calotrophon hystrix García, 2006
Dentimargo gibbus García, 2006
Canalispira aurea García, 2006
Cosmioconcha nana García, 2007
Agathotoma apocrypha (García, 2008)
Darrylia clendenini (García, 2008)
Acmaturris pelicanus García, 2008)
Agathotoma ecthymata García, 2008
Viridrillia aureofasciata García, 2008
Retimohnia acadiana García, 2008
Epitonium pigrum García, 2011
Epitonium leali García, 2011
Isara ulala García, 2011
Falsuszafrona tortugana García, 2015
Latiromitra niveobabelis García, 2015
Falsuszafrona yucatecana García, 2023
Corinnaeturris humilis García, 2024
Glyphostoma coronaseminale García, 2015

Citations to the original descriptions of these species are found in the References. Some 600 species of mollusks dredged by the *Pelican* offshore Louisiana are listed, and many photographed, at https://www.jaxshells.org/efg1030.htm

Another important source of GoMx material came from the Thomas and Barbara McGinn collection, of *Scaphella mcginnorum* fame (see García and Emerson, 1987; García, 1988, 1989). Tom was a shrimper from south Louisiana who trawled in the GoMx for many years. His shrimping area not only covered the northern and eastern Gulf but also the northern Yucatán Peninsula; back then in the 1960s, Mexico allowed such operations. Tom was enthusiastic and careful about his collecting, but Barbara was the "curator" and took good care of the specimens once Tom brought them home. After Barbara died, I had the opportunity to obtain from the family many interesting GoMx species, and have catalogued nearly 200 lots trawled and dredged by Tom. Yes, Barbara had asked him to make a small dredge for the sole purpose of collecting shells.

My other important source of GoMx shells came from Frank Frumar, an enthusiastic shell collector who took upon himself to build several types of dredges to use in deep water off the Florida Keys. Not only did Frank dredge such rarities as *Babelomurex fax*, *Adelphotectonica uruguaya* and *Sthenorytis pernobilis* but also a number of species that now carry his name and that of the skillful captain of the boat that did the dredging, Steve Kern (see García, 2024c):

Calliostoma frumari García, 2007
Astyris frumarkernorum García, 2009
Dilemma frumarkernorum Leal, 2000
Bartschia frumari García, 2008
Prunum frumari Petuch, 2012

The number of GOMx lots in the collection from these three sources total 5559.

****

Some Youtube videos of the collecting expeditions aboard the Pelican in the Gulf of Mexico can be found on https://www.youtube.com/@nemastoma2 in the playlists

Seaweeds and invertebrates in the Gulf of Mexico

https://www.youtube.com/watch?v=b_r6f9dpJKU&list=PL0D8228F7743D514E

and

Oil spill-impacted Gulf of Mexico Seaweeds

https://www.youtube.com/playlist?list=PL7779554C0963D35B

Dredging in the Gulf of Mexico

G- Areas where the R/V Pelican dredged in the Gulf of Mexico (highlighted in yellow). H-,The R/V Pelican. I- After a haul with the box dredge. L--Picking through a dredge haul at night. K- The Benthic Skimmer. L- Collection after a haul with the Benthic Skimmer.

PLATES

1-Cornusepta acuminata, 3.2mm (17022). 2- Diodora arcuata, 18.3 mm (31085). 3- Diodora fluviana, 17.8 ,, (30840).
4- Emarginula phrixodes, 7 mm (32346) . 5- Emarginula sicula, , 7 mm ((31085). 6- Laevinesta atlantica, 5.6 mm (23440).
7- Emarginula tuberculosa, 8.8 mm (24458). 8 Rimula dorriae, 6 mm (16910). 9- Lucapinella limatula, 15.3 mm (26073).
10- Lucapina aegis, 29.4 mm (31960). 11- Rimula frenulata, 4.3 mm (26503). 12- Zeidora bigelowi, 3.3 mm (22246).

13- Haliotis pourtalesii, 32.6 mm (32260). 14- Anatoma crispata, 2 mm (25299). 15- Anatoma proxima, 2.8 mm, (24449). 16- Entemnotrochus adansonianus, 101.2 mm (32299). 17- Kanola meroglypta, 21 mm (30116).18-Perotrochus amabilis, 55 mm,)22167).19- Euchelus guttarosea, 3 mm (32386). 20- Mirachelus clinocnemus, 2.9 mm (32347). 21-Seguenzia hapala, 25310). 22- Sequnzia cf. formosa, 3 mm (25090). 23- Calliotropis globosa, 11 mm (27388). 24- Calliotropis tiara, 4.5 mm (26819).

25- Arene venustula, 7.5 mm (22229). 26- Calliostoma barbouri, 23.5 mm (25218).
27- Calliostoma mariae, 30.7 mm ((32806). 28- Calliostoma benedicti. 32 mm (28881)
29-Calliostoma bigelowi, 28.7 mm (32792). 30- Calliostoma dentatum. 4.8 mm (26442).
31- Calliostoma euglyptum, 7 mm (26426). 32- Calliostoma fascinans, 6.9 mm (26434).
33- Calliostoma frumari, 7.7 mm, Holotype (paratype: 28090).

34- Calliostoma hendersoni, 25.7 mm (32791). 35- Calliostoma cf. hirtum, 4.6 mm (25320)
36- Calliostoma oregon, 17.2 mm (30841). 37- Calliostoma orion, 22.2 mm (25781)
38- Calliostoma psyche, 27.2 mm (2544). . 39- Calliostoma roseolum, 12.3 mm (31248).
40- Calliostoma rosewateri, 34 mm (5203). 41- Lithopoma tectum, 38 mm (28651).
42- Turbo castanea, 34.6 mm (30456)/

43- Turbo caillettii; (L) 33.8 mm 22130);(R)32.3mm (28651). 44-Turbo canaliculatus, 58.9 mm (7325). 45- Turritella lyonsi, 17.8 mm (25811). 46- Carithioclava garciai, 91.6 mm (9588). 47-Propustularia surinamensis, 33.7 mm (10260). 48- Pseudozonaria aequinoctialis, 38.1 mm (6661). 49- Muracypraea mus bicornis 74.6 mm (6259). 50-Callistocypraea broderipii, 100.4 mm (15844)/.

51- Erronea nimphae, 50 mm (15835). 52- Cyphoma intermedium, 31,4 mm (7569).
53- Primovula vanhhyningi, 9.5 mm (30109). 54- Phenacovilva piragua, 13.7 mm (30108).
55- Simnia bartschi, 11 mm (26310). 56- Hipponix cf. benthophila, 8 mm (25514)
57- Natica perlineata, 14 mm (27685). 58- Polinices leptaleus, 14 mm (30113). 59-Echinophoria coronadoi, 103 mm (30887). 60- Oocorys clericus, 108.7 mm (32830). 61- Sigatica carolinensis, 32 mm (25455). 62- Oocorys sulcata, 32 mm (30827). 63-Sconsia nephele, 14 mm (32854).

64- Sconsia grayi, 36.4m (28618). 65- Eucorys bartschi, 122 mm (32847). 66- Tritonoranella ranelloides, 43.8 mm (32000). 67- Bursa ranelloides, 50 mm(25759). 68- Lampasopsis thomae, 22.9 mm (36483). 69- Bursa natalensis, 75.5 mm (31169). 70- Sassia epitrema, 24.5 mm (16896). 71- Distorsio clathrata, 36 mm (30184). 72- Distorsio constricta, 26.9 mm (30196). 73- Distorsio perdistorta, 38.7 mm (30310. 74-Personopsis grasi, 16 mm (32297). 75- Ranularia rehderi, 53.5 mm (31984).

76- Monoplex aquatilis, 124.6 mm (8461). 77- Monoplex krebsii, 92.5 mm (6708). 78- Parviturboides interruptus, 1.5 mm (2449). 79- Parviturboides sp., 2 mm (16130). 80- Niso hendersoni, 35 mm (25216) 81- Niso tricolor, 25,8 mm(33894). 82- Doluchupis leei, 6.5 mm (33249). 83- Niveria harriettae, 14 mm (26053) 84- Pseudopusula antillarum, 4.7 mm(26253). 8%- Serpulorbis decussatus, 13.6 mm(diameter of aperture)((31954). 86- Metaxia spinosai, 3.6 mm (24599). 87- Adelphotectonica urugaya, 20.7 mm (33099).

88-Discotectonica discus, 27.8 mm (33898). 89- Spirolaxis centrifuga, 3.3 mm(24516). 90- Claviscala garciai, 62.7 mm (13247). 91-Epitonium thorssoni, 3.2 mm (32532). 92- Cycloscala montrouzieri, 1.5 mm (29891). 93- Cylindriscala humerosa, 12.9 mm (30383). 94- Gregorioiscala sp., 4.8 mm (30381). 95- Papuliscala cf. japonica, 4.5 mm (30388) 96- Murdochella sp., 3 mm (30382). 97- Variciscala raricosta, 25.3 mm (28956). 98- Amaea retifera, 32.2 mm(32823). 99- Amaea mitchelli, 48.2 mm (7934). 100- Cirsotrema cf. dalli, 10.6 mm (33245). 101- Cirsotrema dalli, 25.4 mm(12952). 102- Cirsotrema pilsbryi, 8.4 mm (24383). 103- Cycloscala cf. echinaticosta, 6 mm (23319). 104- Cylindriscala rosenbergi, 6.8 mm (11156).

105- Eccliseogyra formosissima, 19 mm (105). 106- Epitonium celesti, 10.1 mm (33315) 107- Epitonium krebsii, 27.7 mm (30001). 108- Epitonium leali, 4.3 mm (12861). 109-Epitonium pigrum, 11 mm (26207). 110- Gregorioiscala pachya, 16 mm (27415). 111- Gregorioiscala abbotti, 2.6z mm (7839). 112- Opalia aurifila, 5 mm (26291). 113- Opalia burryi, 14 mm (27750). 114- Opalia eolis, 4.3 mm (25408).115- Opalia morchiana, 5.7 mm 26283).

116- Opalia hotessieriana, 7 mm (30526). 117- Amaea guineensis, 72.3 mm (16136)
118- Amaea bruneopicta, 52 mm (24108). 119- Amaea tehuanarum, 55.1mm (17854).
120-Cirsotrema skoglundae, 12.2mm (29646). 121-Cirsotrema hertzae, 25 mm (18571).
122- Sthenorytis turbinum, 46.8 mm (12610)123- Sthenorytis pernobilis, 38.8 mm (33107)
124- Punctiscala watsoni, 7 mm (33256). 125- Ameranna florida, 11.7 mm (3044).
126- Bartschia frumari, 24.5 mm (31260). 127- Engina itzammai, 17.4 mm (25800).
128- Gaillea canetae, 74.7mm (30433).129 Gailea canetae, 65 mm (27397).
130- Monostiolum harryleei, 19 mm (25819) Holotype.

131- Antillophos smithi, 32.9 mm (31429).132- Busycoarctum coarctatum, 30.2 mm (26098). 133- Aristofusus halistreptus, 37.8 mm(31974).134-Aristofusus excavatus, 65.6 mm (30793). 135- Cinctura hunteria apalachee, 90.9 mm(4095). 136- Cinctura branhamae, 77.3 mm)8718) 137- Dolicholatirus paulii, 21.7 mm (33082). 138- Fasciolaria bullisi, 149.3 mm (33053) 139- Fusinus aepynotus, 31.6 mm (31944). 140- Fusinus benthalis, 15 mm (9265). 141- Fusinus colombiensis, 37.9 mm (5929). 142- Heilprinia couei, 73.5 mm (25982). 143- Heilprinia timessa, 108.3 (28812). 144- Heilprinia sp.,, 50.3 mm (27599).

145- Polygona nemata, 30 mm (27968). 146- Fusinus bleakeanus, 40.1 mm (31730). 147- Fasciolaria cf. tephrina, 230 mm (30890). 148- Fasciolaria tephrina, 220 mm (31167). 149- Fasciolaria delicatissima, 169.9 mm (31404).150- Heilprinia dowiana, 72.5 mm (5314). 151- Fusilaria garciai, 225 mm (31106). 152- Heilprinia coltrorum, 113.8 mm (31323).

153- Columbella marrae, 7.3 mm (31557). 154- Astyris diaphana, 8.7 mm (25228)
155- Astyris frumarkernorum, 4.8 mm (30063). 156- Astyris verrilli, 8.4 mm (24452).
157- Astyris profundi, 5 mm (25176). 158- Cosmioconcha geigeri, 6 mm (26311).
159- Cosmioconcha nana, 4.6 mm (27873). 160- Cosmioconcha aff. rikae, 16.7 mm (30160).
161- Steironepion minus, 4.4 mm (14727). 162- Falsuszafrona tortugana, 8.8 mm (31234).
163- Falsuszafrona yucatecana, 10 mm, holotype (ANSP 493838).

164- Colubraria procera, 66.1mm (8736). 165- Morum strombiformis, 28.8 mm(12056). 166-Morum dennisoni, 53.7 166mm (30104)167- Harpa gracilis, 26.9 mm (10399). 168- Harpa goodwini, 70.2 mm (32052). 169- Favartia hidalgoi, 26.2 mm (28604). 170- Pygmaepteris richardbinghami, 14.7 mm (32261). 171- Babelomurex dalli, 18.4 mm (25048). 172- Boreotrophon lacunellus, 7.2 mm (24929). 173- Calotrophon hystrix, 28.8 mm, holotype (25831
173- Calotrophon carnicolor, 25.8 mm (18504).

175- Coralliophila aedonia, 17.5 mm(25047). 176- Coralliophila cf. macleani, 14.6 mm (13915)
177- Coralliophila richardi, 11.3 mm (22730). 178- Coralliophila cf. richardi, 14 mm (14594)
179-Coralliophila sp., 29.6 mm (5355). 180- Dermomurex cf. paupercula, 14.2 mm (27831)
181- Favartia cellulosa, 19 mm (23321). 182- Favartia minirosea, 8.5 mm (25854).
183- Murexul emipowlusi, 6.5 mm (14569) 184- 184- Favartia taylorae, 21.5 mm (25817).
185- Hirtomurex cf. subsquamosus(juvenile), 5 mm (24930).

186- Paziella nuttingi, 46.1 mm (23912). 187- Paziella pazi, 38.6 mm (31154)
188- Pazinotus bowdenensis, 10.2 mm (30429). 189- Chicoreus bullisi, 65.5 mm (31124)
!90- Chicoreus mergus,,71.5 mm (5586).191- Favartia puntagordana,17.2 mm (19229)
192- Murexul sunderlandi, 24.6 mm (31425) 193- Paziella oregonia, 58 m (9534).
194 Poirieria actinophora, 16 mm(18505).195- Pygmaepterys germainae, 6.8 mm (18482).
196- Pygmaepterys lourdesae, 5.1 mm (31030)186

197- Laevityphis bullisi, 31.1 mm (10202). 198- Siphonochelus longicornis, 16.1 mm (31141)
199- Tripterotyphis triangularis, 20 mm (11679). 200- Vokesimurex cabritii, 62 mm (26022)
201 Trophonella longstaffi, 38.3 mm (9716). 202-- Trophonella shakletoni, 39.4 mm (10149)
203- Siratus gundlachi, 66.2 mm (31321). 204- Siratus vokesorum, 32 mm,(18491) Holotype
205- Siratus aguayoi, 49.5 mm(18503). 206- Siratus tenuisculptus, 61,7 mm (9206)

207- Siratus beauii, 123.6 mm (30829). 208- Siratus beauii, 141.8 mm (32784) 209- Siratus beauii, 134 mm (18507) 210- Vokesimurex anniae, 30.7 mm (25844) 211- Siratus consuela,, 52,2 mm(L),51 mm(R) (32305) 212-Siratus anniae, 37 mm (25846). 213- Vokesimurex tryoni, 44.2 mm (18518). 214- Cinclidotyphis myrae, 19.3 mm (28422). 215- Vokesimurex sallasi, 43.4 mm (31963)..

216- Phyllonotus mexicanus, 73.1 mm,(32285). 217- Phyllonotus mexicanus, 117.3 mm (7939)
218- Typhinellus labiatus, 19.5 mm (26267). 219- Typhina lamyi, 15.8 mm (23262)
220- Timbellus havanensis, 18.9 mm (33238). 221- Austrotrophon catalinensis, 68.4 mm (28814)
222- Favartia dipsacus, 32.4 mm (24756). 223- Paziella galapagana, 46.1 mm (18508).
224- Latiaxis santacruzensis, 48.9 mm (12611).

225- Pterotyphis fimbriatus, 28.2 mm (23038) 226- Purpurellus macleani, 32.9 mm (22272)
227- Zacatrophon skoglundae, 82.4 mm (10347) 228- Latiromitra niveobabelis, 45.9 mm (31199)
229- Vexillum styria, 14 mm (25457) 230- Vexillum pulchellum, 16.9 mm (22165)
231- Vexillum articulatum, 22.4 mm (30123) 232- Vexillum hendersoni, 18.5 mm (26222)
233- Vexillum garciai, 5 mm (26309).

234- Vexillum sikesi, 9.3 mm (31071) 235- Siphovasum latiriforme, 50 mm (17336) 236- Volutomitra bairdii, (L) 39.4 mm, (R) 29.6 mm (30178) 237- Neotiara gausapata, 8.8 mm (30632) 238- Panamitra gigantea,, 92.4 mm, (766)) 239- Isara straminea, 28.3 mm (23154) 240- Subcancilla sp. aff. candida, 30.4 mm (25889) 241- Isara ulala, 22.4 mm (32286) 242- Isara ulala, 28.4, (32316). 243- Dibaphmitra florida, 74.4 mm (5639)

244- Fusimorio deshayesi, 32 mm (16139) 245- Euclia laurettae, 41 mm, (20152) 246- Trigonostoma bullatum, 74.1 mm (22583) 247- Cancellaria rosewateri, 16.3 mm (16886) 248- Ventrilia jonathoni, 35.1 mm (32846) 249- Ventrilia tenera, 25.3 mm (31094) 250- Tritonoharpa sp. aff lanceolata, 12.5 mm (30516) 251- Dentimargo hennequini, 5 mm (26252) 252- Dentimargo gibbus, 5.6 mm (26345) 253- Prunum frumari, 12.9 mm (30823)

254- Prunum cassis, 23.3 mm (32807) 255- Prunum sp., 10.8 mm (32334)
256- Prunum rostratum, 11.53 mm (26007) 257-Volvarina aldeynzerae, 8.5 mm (27904)
258-Volvarina maya, 8 mm (26244) 259-Jaspidella miris, 6.1 mm (26385)
260- Oliva circinata, 40.5 mm (25786) 261- Lyria beauii, 67 mm (8818)
262- Scaphella cuba, , 43.6 mm ((22756).

263- 268 (4083, 2975, 4965, 3818, 3418)
Aurinia kieneri- All of these specimens were trawled between Horn Island, Mississippi and SW Pass, Mississippi Delta,from72 m to130 m, in gray mud Notice top (right) specimen without the brown patter for a period of its growth.

269- Aurinia kieneri, 162 mm (4965) 270- Aurinia kieneri, 35.3 mm (30786) 271- Auriniadubia dubia, 70.5 mm (22754) 272- Auriniadubia georgiana, 70.5 mm (22754) 273- Aurinia dubia georgiana, 83.1 mm (33043) 274- Aurinia dubia dubia, 72.2 mm (32827) 275- Caricellopsis contoyensis, 160 mm (7926) 276- Aurinia dubia schmitti, 93.7 mm (3115 8) 277- Aurinia dubia schmitti, 74.1 mm (33039).

278- Aurinia kieneri, 140 mm (30847) 279- Scaphella dohrni, 71.4 mm (2895) 280- Scaphella robusta, 125.2 mm (32775) 281- Scaphella robusta, 106.8 mm (7941) 282- Scaphella robusta 62.2 mm (6484) 283- Scaphella worki, 45.9 mm (30830) 284- Scaphella marionae, 67.6 mm (4941) 285- Scaphella marionae, 62.8 mm (4941) 286- Scaphella gouldiana, 79 mm (33049).

287- Scaphella gouldiana, 52.1 mm (15618) 288- Scaphella gouldiana,49.5 mm (33044)
289- Scaphella cf. gouldiana, 45.1 mm (33042) 290- 293-Voluta polypleura f. kotorai,
290-92.9 mm (5583) 291- 107.2 mm (31893) 292- 39 mm (adult) (15820) 293- 62.3 mm (6260)
294- Voluta polypleura cross of formslhilli and kotorai, 78.2 mm (6260)

295- Voluta polypleura, 58 mm (10722) 296- Voluta polypleura f. garciai, 84 mm (9119) 297- Voluta polypleura, 80.7 mm (9540) 298- Voluta polypleura f. demarcoi, 86.8 mm (5392) 299- Voluta polypleura f. hilli (?) 93.2 mm (9036) 300- Voluta polypleura f. hilli, 95mmm (5583) 301- Voluta polypleura f (?), 89 mm (31892) 302- Voluta polypleura f. hilli, 89.4 mm (31892) 303- Voluta polypleura f. hilli,78.6 mm (5393) 304- Voluta polypleura f. hilli, 66.5 mm (11055) 305-Voluta polypleura f. demarcoi(?), 76.5 mm (33030)

306- Voluta polypleura f. hilli, 69.6 mm (8933) 307- Voluta polypleura f. sunderlandi, 50 mm (9351) 308- Voluta polypleura f. sunderlandi, 51.5 mm (10730) 309- Voluta polypleura form, 59 mm (16089) 310-Voluta polypleura form, 59.8 mm (31300) 311- Voluta polypleura form, 73.1 mm (31301) 312- Voluta polypleura form, 55.2 mm (31302) 313- Voluta polypleura form, 70 mm(32638) 314- Voluta polypleura form, 63.5 mm (30921) 315-317- Voluta polypleura retemirabilis, (L-R) 66.4 mm, 68.8 mm, 76.7 mm (13041).

318-324- Voluta morrisoni, (L to R) 72.5 mm, 56.4 mm, 67.8 mm, 48.4 mm, 60.3 mm,43.3 mm. 325-326- Voluta morrisoni, 50.2 mm, 55.3 mm (11054) 327- 36.5 mm, (32643) 328- Voluta virescens, 101.8 mm (6025) 329-330- Voluta virescens, 78.5 mm, 75.2 mm.

331- Voluta lacertina, 37.9 mm (4901) 332- Voluta lacertina, 56.2 mm (8712) 333- Voluta musica thiarella, 70.6 mm (4781)334- Voluta m. musica, 70.6 mm (4781) 335- Voluta musica, 45.2 mm (5231) 336- Voluta musica, 45.5 mm (6207) 337- Voluta m. musica, 91.6 mm (10256) 338- Voluta musica thiarella, 73.4 mm (5229) 339- Voluta musica thiarella, 66.4 mm (10980) 340- Muluta musica carneolata, 43.7 mm (8865) 341- Voluta musica carneolata, 55.4 mm (9162) 342- Voluta musica carneolata, 54 mm (31322) 343- Voluta musica guinaica, 63.3 mm (3535) 344- Voluta musica guinaica. 68.4 mm (19189).

345-346- Voluta musica thiarella, (L to R) 71.1 mm, 60.8 mm (4780) 347- Scaphella garciai, 41.5 mm (32777) 348- Conus cf. kohni,(22584) 349- Conasprella armiger, 34.3 mm (30878) 350- Conasprella mindana, 38 mm (23164) 351-Conus glenni, 16.9 mm (25656) 252- Conus ritae, 24.9 mm (25559) 353- Conus sahlbergi,(L to R) 15.5mm, 18 mm (15822) 354- Conus kulkulkan (L to R) 34.4mm, 34.3mm, 37.3 mm (10262) 355- Conus harlandi, 63.4 mm (30884).

356a- Conasprella sauros, 34.2 mm (32810) 356 b- Conasprella sauros, 17.9 mm (25813) 357- Conus aureonimbosus, 34.3 mm (30446) 358- Conus cancellatus (from L) 55.5 mm, 44 mm, 21.6 mm, 57.5 mm; all from Louisiana. 359- Conus ermineus, largest, 91.2 mm (6554); all from Louisiana. 360- Conus largillierti (top) 43.8 mm; (bottom) 43 mm (32800).

361- Conus mcgintyi, 36.2 mm (30882) 362- Conus mus, 32 mm (6951) 363- Conus patae, 18.9 mm (25804) 364- Conus patae, 10.7 mm (25830) 365- Conasprella rainesae, 26.1 mm (31197) 366- Conasprella rainesae (a-b) 18 mm(25791) 367- Conus villepinii, 47 mm (3088) 368- Conus granulatus, 48.3 mm (10008) 369- Conus granulatus (L to R) 51.4 mm, 55.5 mm, 56.2 mm (12427).

370- Conus lorenzianus (L to R) 60.1 mm, 65.2 mm (7913) 371- Conus lorenzianus, 115.8 mm (9698) 372- Conus lorenzianus (L to R) 63.1 mm, 73.3 mm, 55.9 mm, (6026) 373- Conus lorenzianus (ridged), 58.9 mm (7910) 374- Conus lorenzianus, 61.5 mm(7938) 375- Conus therriaulti, 72.2 mm (15408) 376- Conus c f. spurius (concave spire), 87.6 mm (7838) 377- Conus spurius, 53 mm (10259)

378- Conus spurius, 57.2 mm (8932) 379- Conus spurius, 71.2 mm (12042) 380- Conus attenuatus, 35.8 mm (25498) 381- Conus attenuatus, 14.3 mm (30393) 382-344- Conus spp. aff. daucus, (L to) 32.7 mm(22113), 38 mm (22146), 47 mm (30102) 385-387- Conus sp. aff. daucus, (L to R), 23.6 mm (31966), 28.2 mm (25787), 25 mm (31958) 288- Conus julieandreae, 28.5 mm (19905).

389- Cryptogemma periscelida, 41 mm (30125) 390- Bathytoma viabrunnea, 45.4 mm (31200) 391- Retidrillia pruina, 28.3 mm (30871) 392- Typhlomangelia nivalis, 35 mm (28783) 393- Typhlomangelia nivalis, 18.1 mm (31266) 394- Bathyferula garciai, 70.4 mm (31946)holotype. 395- Corinnaeturris humilis, 4.5 mm (25304) 396- Glyphostoma coronaseminale,23.2 mm (31264) 397- Glyphostoma golfoyaquense, 14.5 mm (11612) 398-Glyphostoma dentiferum, 12 mm (14541) 399- Glyphostoma pilsbryi, 14.3 mm (26513).

400- Globidrillia sp. aff. smirna, 20 mm (30156) 401- Globidrillia smirna, 22.5 mm (27414) 402- Bellaspira aurantiaca, 10 mm (24651) 403- Clathrodrillia garciai, 29 mm, (31087) 404- Clathrodrillia wolfei ,12.9 mm (30791) 405- Fenimorea abscondita, (L to R) 35.1mm, 27 mm (12953) 406- Fenimorea kathyae, 28 mm (24379) 407- Fenimorea sunderlandi, 48.1 mm (25835) 408- Cerodrillia occidua, 12.8 mm (26526)

409- Leptadrillia campechiensis, 12.8 mm (26636) 410- Leptadrillia lucaya, 18.3 mm (32384)
411- Spirotropis centimata, 28.7 mm (27392) 412- Splendrillia grandis, 18.4 mm (26218)
413- D arrylia clendenini, 8.2 mm (26633) 414- Darrylia harryleei, 6.6 mm (12518) 415- Leucosyrinx sigsbei, 23 mm (27396) 416- Leucosyrinx subgrandifera, 23.8 mm (31202)
417- Leucosyrinx verrillii, 21 mm (27390) 418- Lyoglyphostoma aguadillanum, 11.2 mm (26507)
419- Lyoglyphostoma hendersoni, 11.6 mm (14893) 420- Lyoglyphostoma antillarum, 8 mm (23488)

421- Miraclathurella herminea, 13.7 mm (26509) 422- Eubela mcgintyi, 4.4 mm (26570) 423- Gymnobela edgariana, 75.2 mm (5460) 424- Gymnobela blakeana, 8 mm (24440) 425- Pleurotomella pandionis, 42.1 mm (30010)426- Benthomangelia antonia, 15.5 mm (30486) 428- Cochlespira cavalier, 28 mm (33097) 429- Cochlespira elegans, 40.4mm (4647) 430- Cochlespira laurettamarrae, 57 mm (29641) 431- Cochlespira bevdynzerae, 40.2 mm (31396).

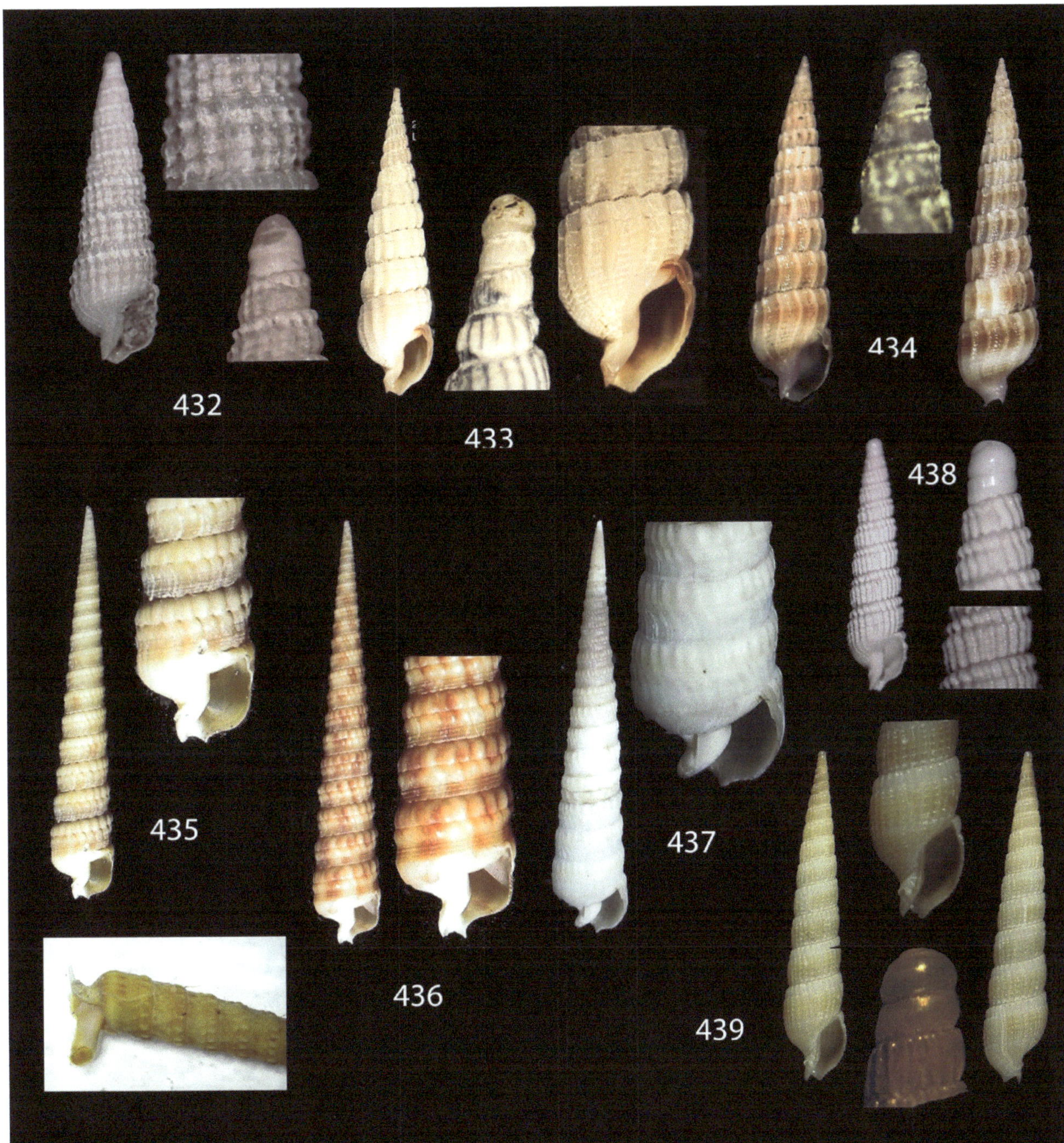

432- Neoterebra acrior, 7.8 mm(14597) 433- Neoterebra arcas, 23.6 mm (30174) 434- Noeterebra mugridgeae, 22.6 mm (31069) 435-436- Terebra lindae 435, 41.4 mm (14093) 436-52.3 mm (27622) 437- Terebra floridana, 69.9 mm (6100) 438- Neoterebra limatula, 8.7 mm (14650) 439- Neoterebra nassula, 22.3 mm (30775).

440- Bullina exquisita, 9.6 mm (28589) 441- Bulla solida, 50.1 mm (30414) 442- Crenilabium exile, 14.7 mm (17155) 443-449 - Specimens from hydrocarbon cold seeps. 443-Cantranea mcleani, 16.3 mm (24041) 444 - Kanola meroglypta, 16.4 mm (24124) 445 - Eccliseogyra formosissima, 23.2 mm (24043) 446 - Retimohnia acadiana. 20.9 mm (24137) 447 - Miowateria malmii, 3.3 mm (24054) 448 - Jorgenia louisiana 53.3 mm (24040) 449- Lucinoma atlantis, 53.2 mm (24039)

450-Hyotissa mcgintyi, 44.4 mm (30422) 451 - Neopygnodonte cochlear, 33.7 mm (29974) 452- Atrina sp. aff. seminuda, 250 mm (29973) 453- Amygdalum polytum, 59.4 mm (13624) 454 Ctenoides sanctipauli, 24 mm (27385) 455- Acesta oophaga, 112.4 mm (32468)

456- A equipecten glyptus, 51.3 mm (29983) 457- 460- Antillipecten antillarum:
457-14.4 mm(9180) 458- 19,9 mm (25486) 459- 25.4 mm (25485) 460- 10.7 mm (26160)
461- Pecten chazaliei, 31.2 mm (24922) 462- Cryptopecten phrygium, 38.6 mm (33078)
463- Cryptopecten phrygium, 44 mm (26225) 464- Laevichlamys multisquamata,
40 mm (26084)

465- Cyclopecten leptaleus, 6 mm (32359) 467- Delectopecten vitreus, 20 mm (16921) 467- Propeamussium simoni, top 15.7 mm, bottom 13.6 mm (29972) 468- Nodipecten fragosus, 90.3 mm, 82.2 mm, 79.2 mm (17030) 469- Nodipecten fragosus, 81.8 mm (20119)

470- Nodipecten fragosus, 160 mm (32782) 471- Nodipecten fragosus, 146 mm (33024) 472- Nodipecten fragosus (L to R) 73 mm, 62 mm, 59.5 mm (17030) 473- Nodipecten nodosus, (L to R) 51.4 mm, 46.9 mm (8841) 474- Nodipecten nodosus, 54 mm (5312)

475- Euvola laurentii, 56.5 to 82.7 mm (5587) 476- Adamussium colbecki, 66.9 mm (17309) 477- Propeamussium simoni, 13.6 mm (29972),,478- Propeamussium dalli, 62.2 mm, (13768).

479- Spondylus americanus collected by SCUBA diver off Louisiana while cleaning oil rig structures.. Notice the flat lower valve of all the specimens.

479-Spondylus americanus: top, 230 mm; bottom, 225 mm.
Both measured from end of spines. (4112)

479- Spondylus americanus: top to bottom, 185 mm, 195 mm, 118 mm- All measured from end of spines. (4112)

Spondylus tenuis. 480- (L to R) 81.3mm, 83.3 mm, 99 mm (10130)
481- 100.2 mm (13678) 482- 101.7 mm (8493) 483- 65.3 mm (2985)

484- Papyridea lata, 30 mm (26217) 485- Glossocardia agassizi, 31 mm (30008)
486- Afrolucina lens, 13.5 mm (13649) 487- Calyptogena ponderosa, 94.7 mm (28667)
488- Ferrocina garciai, 11.7 mm (30783) 489- Eulopia sagrinata, 13.5 mm (32362)
490- Lucinoma filosa, 17.4 mm (31267) 491- Lucinoma louisiana, 53.3 mm (24040)

492- Lucinoma luteophila, 22.9 mm, paratype (27398) 493- Cardiomya ornatissima, 7.8 mm (27896) 494- Cuspidaria rostrata, 17.8 mm (30434) 495- Cuspidaria obesa, 10 mm (30163) 496- Cuspidaria microrhyna, 31.3 mm (31284) 297- Cetoconcha margarita, 8.8 mm (27416).

498- Dilemma frumarkernorum, 15.4 mm (33088) 499- Haliris fischeriana, 5.2 mm (25279)
500- Spinosipella agnes, 10.4 mm (26772) 501- Spinosipella aff. agnes, 7 mm (27889)
502- Periploma orbiculare, 28.9 mm (30771) 503- Poromya rostrata, 10 mm (26809)

COMMENTS ON SOME OF THE FIGURES

13- *Haliotis pourtalesii* is very uncommon as empty shells in the calcareous rubble on the banks off Louisiana; it is very rare as a living specimen. It has been found live in 54-95 m depth. A large, empty specimen, measuring 41.5 mm, was dredged in 87-100m of water.

29- *Calliostoma bigelowi*, obtained from the McGinn family, was of great surprise to me, not only because of its rarity but because it was collected in the northern GoMx, off Alabama. The McGinns had several live-collected specimens of perfect quality and with the original label. At first, the finding seems an oddity, but other *Calliostoma* species such as *C. fascinans, C. roseolum,* and *C. marionae* have also been dredged in this area. Species of Ovulidae, another family of hard-bottom preference, have been dredged by the *Pelican* off Alabama; so, there are pockets of hard reef that such families can inhabit.

43- *Turbo cailletii* used to be considered a rare species, but large-size, colorful specimens were rather commonly found on Louisiana offshore calcareous banks.

46- *Cerithioclava garciai* was an astonishing find. The one specimen that was first offered to me I thought to be a joke. I was not acquainted with fossil shells and I was sure it was an Indo-Pacific species. Dr. Richard S. Houbrick, the author of the taxon, thought at first that it was a fossil, as the genus *Cerithioclava* was only known from the fossil records. It took me several more trips to Roatán Island, and several more specimens, to convince him otherwise.

47- *Propustularia surinamensis* is shown here only because of its unusual locality. It is only the second specimen I have obtained in more than three decades of collecting from fishermen.

63- *Sconsia nephele* is a rare species found in the southern Lesser Antilles. This juvenile specimen was sent to me by Dale Bittner, of Key West, Florida, who found it while diving in Union Island. It came unidentified, together with a number of other small species, so it was of great surprise to me.

66- *Tritonoranella ranelloides* is very variable in its sculpture. The specimen shown here is the more tubercular form, but specimens with finer sculpture, formerly known as *Bursa ranelloides tenuisculpta* but now considered a form of *ranelloides*, have also been found off Louisiana.

70- *Sassia epitrema* is a rare species that inhabits waters off southern Australia. My friend Bill Clendenin and had gone to visit his good friend, Jack Austin, of Melbourne, Australia, one of the great collectors of South Australian shells. While there, Jack suggested that we visit another. friend, Peter Weiser, a fisherman, who lived in a beautiful rainforest with giant fern trees and lyrebirds. I had brought with me shells to trade, and this specimen was one of the species I got from Peter (see García, 2000b).

74- *Personopsis grasi* was thought to have the southern Caribbean as its northern limit in the western Atlantic. It was not until the *Pelican* campaigns that the species was recorded from the GoMx. It has been found at three locations off the Louisiana coast.

75- *Ranularia rehderi* is another species that had not been reported from the GoMx before the *Pelican* campaigns. We have four records for Louisiana and two for Bahía de Campeche, Mexico; also, two specimens from Honduras.

90- *Claviscala garciai* was incorrectly placed in the genus *Opalia*.

99- *Amaea mitchelli* seems to be an endemic species of the GoMx. The southern Caribbean species identified as "*mitchelli*" is *Amaea africana* (see García, 2022a).

103- *Cycloscala echinaticosta* - The original figure of this species, as well as other known specimens, is tightly coiled, unlike the specimen shown here. This form is found as frequently as the former and may turn out to be one of Möch's synonymized names (see García, 2004).

117- *Amaea guineensis* was purchased in Senegal in 1996 from Marcel Pin, the well-known shell dealer who lived in that country.

120-121- *Cirsotrema skoglundae* and *Cirsotrema hertzae* are two Panamic Province species named after two great shell collectors: Carol Skoglund, an enthusiastic collector from Arizona who specialized in marine shells from the Sea of Cortez, and after whom *Zacatrohon skoglundae* Houart, 2010 was named; and Carole Hertz, for many years editor of the well-known conchological publication *The Festivus* (see García, 2010a).

122- *Sthenorytis turbinum* is a rare, deep-water species that inhabits the waters from Ecuador to the Galápagos islands. The specimen shown here was obtained from the well-known Florida collectors Jo and Rusty Bennett, after whom *Cirsotrema benettorum* was named. Jo and Rusty corresponded frequently with the also well-known shell collectors from the Galápagos Islands, André and Jaqueline de Roy. The Bennetts obtained many other rare species from the de Roys, such as *Babelomurex santacruzensis* (fig. 224) and *Sveltia gladiator*.

130- *Monostiolum harryleei*, from the Campeche Banks, Mexico, is known from only two specimens, the holotype, shown here, and a long-dead paratype (see García, 2022b).

144- *Heilprinia sp.* is similar to *H. couei* but has a rougher sculpture, more angular whorls and a somewhat different protoconch. It may be a variation of the latter.

147-149- *Fasciolaria cf. tephrina, F. tephrina* and *F. delicatissima* are very similar. *F. delicatissima* has a different protococh, different color pattern and different habitat, as it has been trawled by shrimpers in smooth bottom. The *tephrina* "complex" inhabits very rough, rocky bottoms in an area only rather recently worked by fish-trap fishermen. I was not able to discern any difference between the two forms shown here, other than coloration. All the specimens were collected "crabbed" (see García, Lyons and Snyder, 2016).

151- *Fusilaria garciai* inhabits hard, deep-water bottom at the edge of the NE continental shelf of Honduras and the Honduras-Nicaragua border. It has been obtained only rather recently, when Honduran fishermen started setting baited fish traps in that otherwise unexplored habitat. All the specimens were collected "crabbed" from baited traps.

152- *Heilprinia coltrorum* was originally described from "the Florida Straits"; however, this species has not been reported from that area other than in its original descrIption. The specimen shown here and several others were collected in baited fish and lobster traps at the edge of the continental shelf in NE Honduras.

189- *Chicoreus bullisi* is a relatively deep-water species and rarely found in collections. Its type locality is "off the Caribbean coast of Nicaragua". The specimen shown here was collected "crabbed" in 200 m near the type locality.

203- *Siratus gundlachi* had been confused until recently with *S. articulatus*. This specimen, like other rare species treated in this work, was obtained in fish traps set in deep water at the edge of the continental shelf in NE Honduras.

204- *Siratus vokesorum* is known from only two specimens. The holotype, shown here and taken in 273 m of water, was originally in the collection of the well-known muricid collector Eugenia Wright, who deposited it in Dr. Emily Vokes collection at Tulane University, New Orleans.

206- *Siratus tenisculptus* is a deep-water species. The figured specimen was given to me by Don Olsson, a diver who had a small dive shop in La Parguera, southern Puerto Rico. I obtained it from him when I was visiting in 1969.

207-209- *Siratus beauii*, a species that lives in relatively deep water, is variable in its sculpture, from rather plain shells (fig. 208), to shells with extended "wings" (fig. 209). The specimens shown in these two images are what may be considered maximum size. The Louisiana specimen (fig. 207) is of an unusually dark color.

216-217- *Phyllonotus mexicanus* was, until rather recently, confused with *P. pomum*. The specimens shown here represent the northern, Louisiana population (fig. 216), and its Yucatecan counterpart (217).

219- *Typhinellus lamyi*, originally described from the island of Guadeloupe, in the Lesser Antilles, is rare in the GoMX. The specimen shown, a single specimen dredged by the *Pelican*, was collected off the Louisiana coast.

222- *Favartia dipsacus* is a rare Chilean species. This specimen was given to me by the well-known Chilean collector Jorge Basly Santa María, while visiting him in Iquique, Chile. He only had two specimens.

231- *Vexillum articulatum* is a challenging taxon, as it is very similar to other western Atlantic taxa of the Costellariidae. For more information see García, 2011d, 2023a.

236- *Volutomitra bairdii*, although rare in private collections, it is a common species in the northern GoMx off Louisiana, between 680 and 1200 m. It has been dredged alive between 688 and 801 m (see 2011c).

240- *Subcancilla* species. This single specimen was dredged by the *Pelican* in the GoMx, as well as the first specimen of the genus *Subcancilla* dredged in that body of water.
241-242- *Isara ulala* seems to be endemic to the northern GoMx. Although it has been dredged in the banks off Louisiana at 19 stations, it was not collected anywhere in the GoMx where the *Pelican* has dredged (see García, 211a).
244- *Fusimorio deshayesi*. This specimen came from the well-known dealer Marcel Pin, whom I visited while in Senegal.
246- *Trigonostoma bullatum*, a gigantic specimen from Panama Bay.
247- *Cancellaria rosewateri* is represented in my collection by two specimens collected in a bottom consisting almost exclusively of mineral nodules (see García, 1997, 2021).
248- *Ventrillia jonathoni* is a recently described species from northern Yucatán. In its original description it was reported to attain a maximum size of 12 mm. However, this specimen, also from northern Yucatán, measures 35.1 mm.
250- *Tritonoharpa* sp. *aff. lanceolata*, differs from *T. lanceolata* in pattern and sculpture. It is the only specimen in the collection, so it may only be an oddity.
252- *Dentimargo gibbus* is a common species in restricted areas of the Campeche Banks, Mexico, where it was collected at four stations (see García, 2022c).
255- *Prunum species* is closely related to *Prunum cassis*, as it has a rather heavy parietal shield; however, it differs in size and profile (see García, 2022c). It has been dredged by the *Pelican* at 8 stations, from the Florida Panhandle to Louisiana.
263-270- *Scaphella kieneri*. More than 100 specimens of *S. kieneri* have passed through my hands, all collected by a fisherman friend of mine who lives in Alabama and by Tom McGinn (see earlier section on the R/V *Pelican* and the GoMx) They vary in profile from specimens with wide shoulders (fig. 265), which tend to have some slight axial folds, to very slim specimens (fig. 269). Some rare specimens lack the brownish square spots characteristic of the species (fig. 265). Figure 264 shows a specimen that became "spotless" for a while and then returned to its normal pattern at the lip edge. I am treating *Scaphella kieneri* as a separate species from the *S. dubia* complex, which has smaller shells of paler color and different protoconch formation. Although *S. kieneri* has been reported from the eastern part of the Florida panhandle, such localities are doubtful, as this species seems to inhabit very muddy, rather than sandy, bottoms.
283- *Scaphella worki*. Although this taxon has been synonymized with *Scaphella robusta marionae*, it has a narrower shell and does not conform to the *marionae population* from Florida. The specimens dredged by the *Pelican* were all in one haul and very much alike, and may be juveniles of *S. robusta*. However, they are still narrower than *the* latter, so I have arbitrarily opted to consider *S. worki* to be a full species.
284- *Scaphella cf. gouldiana.* I have not been able to discern any differences between the white form shown here from *S. gouldiana, a* species that is rather variable in coloration.

290-314- *Voluta polypleura* complex (see García, 1993). I am showing in these images the remarkable variations that the *Voluta polypleura* complex show in color pattern and, more restricted, in shape. As one moves from bank to bank in NE Honduras and Nicaragua, different color patterns, shapes and sizes are found. See, for example, figures 310, 311 and 312. What would one name them?

I have in my collection slightly over 100 specimens of this complex, and they were kept for specific reasons out of some 300 hundred that I obtained in Roatán Island through the years. The complex varies not only in coloration but also in the size of the protoconchs; they are also variable in the size of the specimen. Figure 292, a fully adult specimen, measures only 39 mm, one of several equally small specimens from the same population.

314- *Voluta polypleura* sp. *aff. V. p. retemirabilis*. I have a number of specimens of this form that, although generally similar to *retemirabilis*, have some distinctive characters. Moreover, they have a different habitat, as they live in deep water at the edge of the continental shelf in NE Honduras. The specimens were collected "crabbed" in fish traps, and may be a separate subspecies of *V. polypleura*.

315-317- *Voluta polypleura retemirabilis*. I consider this to be a valid subspecies not only because of the unique difference in its pattern but also because of its location: Misteriosa Bank, a submerged bank almost equally distant from the Mexico/Belize/Honduras coastlines and west of the Cayman Islands, and therefore rather distant from the eastern Honduras/Nicaragua/Colombia (San Andrés I.) populations. When fresh-collected, this subspecies has the shiniest surface of any other form or species in this complex, with perhaps the exception of *Voluta morrisoni*.

318-327- *Voluta morrisoni*. This species is more conservative in color pattern than *polypleura*. The one striking form is the specimen in fig. 327, of which I have a single specimen.

331- 332- *Voluta lacertina*. This may be the northern form of *V. virescens*; however, I treat it here as a full species because I have not seen any intergrades. It is uncommon in the Honduras area and rare in Belize, which seems to be its northernmost boundary and from where I have obtained only one specimen (fig. 331).

350- *Conasprella mindana*. Although it was not reported from the GoMx before the *Pelican* campaigns, this species has since been dredged at one station in Bahía de Campeche, Mexico and 5 stations off the Louisiana coast (see García, 2023c).

355- *Conus harlandi*. This very large specimen was collected "crabbed" in a lobster trap in a Honduran offshore bank; an unusual locality, and at an unusually deep locality.

356- *Conasprella sauros*- The largest specimen known (fig. 356a) was trawled off Yucatán; the smallest specimen, not too far away in Bahía de Campeche. The latter specimen (fig. 356b) is paratype no. 7; although collected empty, it is in pristine condition (see García, 2023c).

359- *Conus ermineus* is rather commonly dredged in the offshore banks of Louisiana, where it grows very large and with unusually bright colors. The *Pelican* has dredged this species at 16 stations (see García, 2023c).

365- *Conasprella rainesae* has not been collected alive by the *Pelican*; however, a number of empty specimens were collected off the west coast of Florida and in Bahía de Campeche. Several specimens were white in coloration (fig. 366b) and are indistinguishable from *Conasprella pacei* (see García, 2023c).

376- *Conus cf. spurius.* This population from northern Yucatán develops a concave spire as it matures; it is comparatively lighter in weight that *C. spurius*, and grows larger in size than other *spurius* populations in the GoMx; also, the periostracum is darker and thicker than other *spurius* population.

382-387- *Conus* sp. *aff. daucus.* This is a rather variable population both in coloration and in shape. It inhabits the offshore banks off Louisiana and Texas, in calcareous rubble. *Pelican* campaigns have not found this species in the southern GoMx or in western Florida (see García, 2023c).

388- *Conus julieandreae.* Several specimens of this interesting species were dredged live at Isla Providencia, Colombia, an island located in the western Caribbean. I also have specimens from South Key, near the Honduras-Nicaragua border, and from Panamá.

392-393- *Typhlomangelia nivali.* This species had not been reported from the GoMx. *Pelican* dredgings have collected the species off Louisiana and west Florida. The specimen shown is also the maximum reported size reported (see García, 2021b).

396- *Glyphostoma coronaseminale.* The pictured specimen is the holotype; the original description was done from that single specimen. Since then, a number of specimens have come to light. I have 5 lots in the collection, all from southern Florida (see García, 2015c).

414- *Darrylia harryleei.* Because of the similarity of the genera *Darrylia* and *Miraclathurella*, some earlier taxa were incorrectly placed in the latter, such as what are now *known as Darrylia kleinrosa* and *Darrylia clendenini* (fig. 413) *(*see García, 2016a). *Darrylia haryleei* specimens were found at several localities in south-central Roatán Island; however, resort constructions have silted the water of the area and, in spite of my arduous work in the last few trips to the island, I found no specimens, dead or alive.

428-431. *Cochlespira* spp. For more than two decades the genus *Cochlespira* was represented in the western Atlantic by only three species; now the number has doubled. *Cochlespira cavalier* (fig. 428) is a rare species; it seems to be confined to SW Florida. *Cochlespira elegans* (fig. 429) is arguably the most "elegant" of the genus, reaching a size of 69.5 mm (EFG 32783). *Cochlespira bevdeynzerae* is not uncommon in the waters off northern Colombia but has been mistaken with *C. elongata* Simone. And *C. leeana,* similar to *C. radiata* (see García, 2010b) and not shown here, is known from Colombia and Barbados. The Panamic Province has also provided a new impressive species, *C.*

laurettamarrae (fig. 430; see García, 2001)). It is also very rare and, as of this writing, there are only two specimens known.

432- *Neoterebra acrior* has had a controversial history with *N. limatula* (fig. 438). For specific differences see García, 2012.

434- *Neoterebra mugridgeae* is found in very muddy situations off both sides of the Mississippi Delta, in about 18-100 m of water. Its outstanding character is a multispiral protoconch of 4 whorls. The *Pelican* has dredged it at 10 stations but it is known only from this relatively limited area.

435-436- *Terebra lindae*. Described from off Apalachicola, Florida, we also have records for Alabama and Louisiana. The largest specimen measures 100 mm (EFG 32787) (see García, 2012).

452- *Atrina* sp. *aff. seminuda*. This is a large, deep-water *Atrina* that is undescribed. A single specimen was collected off Louisiana between 250 and 334m. Dr. Darryl Felder, the principal investigator in that campaign, mentioned anecdotally to me that he had seen larger specimens on earlier campaigns.

453- *Amygdalum politum* is not uncommon at the appropriate depth. The left figure shows the shell attached to its "nest".

475- *Euvola laurentii*. All the specimens were collected by the same shrimp trawler off NE Honduras.

477- *Propeamussium simoni*. Shown here are the right and left valves of the recently described species. Although a number of specimens were dredged at the station of the shown figure, the type locality, the species was absent elsewhere.

479- 479e- *Spondylus americanus*. Most of the specimens figured in these plates were collected inside oil pipes and other submerged man-made structures off the Louisiana coast; thus, the flat lower valve of the specimens. The shells were collected by Thom Anderson, a professional SCUBA diver whose job was to "clean" such structures. What is lost in ornamentation in the lower valve of this form is gained in the profusion of spines of the top valve.

498- *Dilemma frumarkernorum*. This species, known from only a few specimens, was dredged by Frank Frumar in deep water SW of Key West, Florida. A new genus had to be created for this carnivorous species, which has been placed in the family Poromyidae.

500-501. *Spinosipella* spp.- Previously unreported species from the GoMx (see García, 2008d).

INDEX

Cat. No.	Genus	Species	Author	Year	Locality	Country
2544	Calliostoma	psyche	Dall	1888	Dredged in the Gulf Stream, off Key West, in 80 fms.	United States
2895	Scaphella	dohrni	Sowerby	1903	Dredged in 80 fms. SW of Sand Key	United States
2975	Scaphella	kieneri	Clench	1946	Tr. off Horn I.	United States
2985	Spondylus	tenuis =ictericus	Schreibers	1793	Off Sanibel I.	United States
3451	Scaphella	kieneri	Clench	1946	Tr.in 45 fms. off Southwest Pass, Mississippi River Delta	United States
3535	Voluta	musica guinaica	Lamarck	1758	Amuay, Amuay Bay,Paraguaná Peninsula-	Venezuela
3818	Scaphella	kieneri	Clench	1946	Tr. in 36-40 fms. S of Dauphine I.	United States
4083	Scaphella	kieneri	Clench	1946	Trawled in 50 fathoms off Horn I.-gray clay bottom	United States
4095	Cinctura	hunteria apalachee	Petuch & Berschauer	2020	Trawled off Dauphine I. 30.2543° N, 88.1124° W	United States
4112	Spondylus	americanus	Hermann	1781	S. of Marsh I. on Platform 73 in 95' of water	United States
4647	Cochlespira	elegans	Dall	1881	Dredged in 2l0 fms off Egmont Key	United States
4780	Voluta	musica thiarella	Linnaeus	1758	Off Gros Ilet, N. of Castries	Saint Lucia
4781	Voluta	musica	Linnaeus	1758	Near Ilet du Tresor(Presq'Ile de la Caravelle)	Martinique
4941	Scaphella	robusta	(Dall	1889	Tr. ESE of South Pass, Mississippi River Delta, in 280 fms.	United States
4965	Scaphella	kieneri	Clench	1946	Tr. off mouth of Mississippi River in 65 fms.	United States
5203	Calliostoma	rosewateri	Clench & turner	1960	85 mi. SE of Galeota	Trinidad and Tobag
5229	Voluta	musica thiarellla	Lamarck	1811	Hog I. - in a few inches of water	Grenada
5231	Voluta	musica	Linnaeus	1758	Sunset Shores Beach- 4' in sand	St. Vincent & the Gı
5312	Nodipecten	nodosus	Linnaeus	1758	Honduras shrimping grounds	Honduras
5314	Heilprinia	dowiana	Olsson	1954	Honduras shrimping grounds	Honduras
5355	Coralliophila	species A			11º 38' N, 69º 27' ; in 30 fms.- Station 564l	Venezuela
5392	Voluta	polypleura	Crosse	1876	Tr. in Honduras shrimping grounds	Honduras
5393	Voluta	hilli	(Petuch	1987	Caught in trap in Honduras lobster banks	Honduras
5460	Gymnobela	edgariana	(Dall	1889	Tr. ESE of Southwest Pass, Mississippi River Delta, in 560 m.	United States
5583	Voluta	hilli	(Petuch	1987	Gorda Banks, NE of False Cape In lobster trap	Honduras
5586	Chicoreus	mergus	Vokes	1974	Between Punta Patuca and Punta Castilla	Honduras
5587	Euvola	laurenti	(Gmelin	1791	Between Punta Patuca and Punta Castilla	Honduras
5613	Voluta	musica	Linnaeus	1758	Bonaire- under coral bolder 3m. deep; buried in sand	Netherland Antilles
5639	Dibaphimitra	florida f. janetae	(Petuch	1987	Honduras lobster grounds-in lobster traps, NE Honduras	Honduras
5929	Fusinus	colombiensis	(Snyder & Snyder	1999	"Coquette" Station 287	
6100	Terebra	floridana	(Dall	1889	Tr. off Cabo Catoche, N. Yucatán - in 50-60 fms.	Mexico
6207	Voluta	musica	Linnaeus	1758	Lac Bay, Sorobon-2' in sand	Netherland Antilles
6259	Siphocypraea	mus bicornis	Sowerby	1870	Off Puerto Estrella, Cabo de la Vela-Tr. in 32 fms.	Colombia
6260	Voluta	kotorai	Petuch	1981	Trapped N. of Great Corn I.- in sand	Nicaragua
6484	Scaphella	robusta	(Dall	1889	ESE of South Pass, Mississippi River Delta- in 225-250 fms.	United States
6661	Cypraea	aequinoctialis	Schild.	1933	Gobernadora Island, Montijo Bay- Dredged	Panamá
6708	Monoplex	krebsii	(Mörch	1877	Honduras shrimping grounds	Honduras
6951	Conus	mus	Hwass in Bruguière	1792	Ambergris Cay-under rock at low tide	Belize
7327	Turbo	canaliculatus	Hermann	1781	SCUBA in 30' off Cueros, NW coast of Honduras	Honduras
7661	Panamitra	gigantea	(Reeve	1844	Dredged in 80 m. off Playa	Ecuador
7926	Scaphella	contoyensis	Emerson & Old	1979	Tr. in l00 fms. off Cape Catoche, Yucatán	Mexico
7934	Amaea	mitchelli	(Dall	1896	Tr. off Port Arthur, in 20 fms	United States
7939	Phyllonotus	mexicanus	(Petit de la Saussaye	1852	Tr. in 25-30 fms off Contoy Light, Yucatán	Mexico
7941	Scaphella	robusta	(Dall	1889	Tr. in 280 fms. ESE of Southwest Pass, Mississippi River Delta	United States
8461	Monoplex	aquatile	Emers.& Old	1963	NNW of Cape Gracias a Dios- in l3 fms. inlobster trap	Honduras
8493	Spondylus	tenuis =ictericus	Schreibers	1793	SCUBA, Attached to oil rig	Venezuela
8718	Cinctura	branhamae	(Rehder & Abbott	1951	Tr. NW of Holbox I., Yucatán Peninsula	Mexico
8736	Colubraria	procera	(Sowerby	1832	Bahía de Navidad, Jalisco in 60' at night.	Mexico
8818	Lyria	beauii	(Fisch.& Bern.	1857	IGuadeloupe, n fish trap in 250 m.	Guadeloupe
8841	Nodipecten	nodosus	(Linnaeus	1758	Cayos Viborillas-lobster fisherman, SCUBA	Honduras
8865	Voluta	musica	Linnaeus	1758	l-3', on stones in eel grass?sand-S.Barbados	Barbados
8933	Voluta	polypleura	Crosse	1876	Pedro Banks,off S.Jamaica-in l80' in lobster trap-rubble/sand	Jamaica
9036	Voluta	polypleura	Crosse	1876	Tr.NW of Caratasca Keys	Honduras
9119	Voluta	garciai	(Petuch	1987	Trawled "Nicaragua"	Nicaragua
9161	Voluta	musica	Linnaeus	1758	4' in sand, at night; Pigeon Pt.	Trinidad & Tobago
9180	Antillipecten	anillarum	(Récluz	1853	500 yards W.of entrance to Paso Malo lagoon, Varadero	Cuba
9206	Siratus	tenuisculptus	Houart & Garrigues	2023	In fishtrap 80-l20 fms. off CaboRojoLighthouse	Puerto Rico
9265	Fusinus	benthalis	(Dall	1889	Dredged off Key West in 100 fathoms	United States
9351	Voluta	polypleura	Crosse	1876	Cayos Cochinos	Honduras
9534	Paziella	oregonia	(Bullis	1964	Tr. in l50 fms. off Venezuela-PARATYPE	Venezuela
9540	Voluta	polypleura	Crosse	1876	Tr. in l80', off Monkey Pt.-sand/algae botto	Nicaragua
9588	Cerithioclava	garciai	Houbrick	1986	Gorda Banks-SCUBA by lobster fishermen	Honduras
9716	Trophonella	longstaffi	Smith	1907	McMundo Sound-SCUBA in 20m, under ice	Antarctica
10130	Spondylus	tenuis =ictericus	Schreibers	1793	Ship Shoal Block 332-SCUBA in 289 ft.	United States
10149	Trophonella	shackletoni	(Hedley	1911	Dredged in 300 mts.;S.Georgia Island.	Antarctica

10202	Laevityphis	bullisi	(Gertman	1969	Trawled off Guajira Peninsula in 100 m.	Colombia
10256	Voluta	musica	Linnaeus	1758	Isla Caribe, Chacopata, in l-3 mts.; sand bottom with Thalassia	Venezuela
10260	Propustularia	surinamensis	Perry	1811	Cayos Miisquitos, SCUBA in rocks/sand/grass-lobster fisherman	Nicaragua
10262	Conus	cardinalis f. kulkulkan	Petuch	1980	Sandy Bay, NW Roatán I.-SCUBA at night	Honduras
10347	Zacatrophon	skoglundae	Houart	2010	Off Cedros I., Baja California, in 84 m.	Mexico
10399	Harpa	gracilis	Brod & Sow.	1829	Vairua,Takapoto Atoll, Tuamotu Archipelago.	French Polynesia
10722	Voluta	polypelura	Crosse	1876	Punta Paraíso, NE San Andrés I.-snorkelin in 8-l0', sandy bottom, inside re	Colombia
10730	Voluta	polypelura	Crosse	1876	Bogas Key(local name)Cayos Viborillas, NE Honduras- snorkeling in l0'	Honduras
10980	Voluta	musica thiarella	Lamarck	1811	Pointe Borgesse	Martinique
11055	Voluta	polypleura	Crosse	1876	Trawled SW of Gorda Banks, NE Honduras	Honduras
11522	Calliostoma	rosewateri	Clench & Turner	1960	Dredged in l00 fms. west of Dry Tortugas	United States
11612	Glyphostoma	coronaseminale	García	2015	Dredged in 250' SW of Key West	United States
11679	Tripterotyphis	triangularis	(A.Adams	1855	First Bight, S. Roatán Is.-3', underseveral layers of rubble, inside reef	Honduras
12056	Morum	strombiforme	Reeve	1842	One mile E. of Calabash Bight, S. Roatán-crabbed at shoreline	Honduras
12518	Darryllia	harryleei	García	2008	The Key, Oak Ridge, Roatán Is.-under rock, on reef at exposed low tide	Honduras
12610	Sthenorytis	turbinum	Dall	1908	Dredged in 200 m.;Santa Cruz Is.,Galápagos	Ecuador
12611	Babelomurex	santacruzensis	Emerson & D"Attilio	1970	Dredged in 200m. off Santa Cruz Is,Galápagos.	Ecuador
12861	Epitonium	leali	García	2011	Dredged in 28-38' , Heald Bank, off Galveston- in grunge	United States
12952	Cirsotrema	dalli	Rehder	1945	Dredged at 27º37' N; 83º 13" W in 77 ft. of water	United States
12953	Fenimorea	abscondita	Fallon	2016	27º37' N; 83º 13" W in 77ft., muddy sand & shell rubble	United States
13247	Claviscala	garciai	(Kilburn	1994	Balicasag Is., off Bohol- in tangle nets	Philippine Islands
13624	Amygdalum	politum	(Verrill & Smith	1880	28º 27.01'N, 89º 53.44'W, in 450 meters	United States
13649	"Myrteopsis"	lens	(Verrill & Smith	1880	28º27.01'N, 89º53.44'W in 450 meters-	United States
13678	Spondylus	tenuis =ictericus	Schreibers	1793	In 200-300", Platform 632-A, "Garden Banks"-Pltform in 694"	United States
13768	Propeamussiun	dalli	(E.A.Smith	1886	Trawled in 240 fathoms NW of Tampa by shrimp boats	United States
13915	Coralliophila	cf.macleani	Shasky	1987	Dredged at 11º38'N; 69º27'W in 30 fms - R/V "Oregon" Station 5641	Venezuela
14093	Terebra	lindae	Petuch	1987	Dredged 27º48.90"N.; 84º87.01"W.-in 90 meters	United States
14405	Epitonium	celesti =pourtalesii Vei	(Aradas	1854	Dredged at 29º20.25'N.; 87º37' W.-in 250 m.	United States
14541	Glyphostoma	dentiferum	Gabb	1873	29º21' N., 87º42' W., in 140 meters	United States
14569	Murexsul	emipowlusi	(Abbott	1954	29º21'N., 87º42'W; in 140 meters-	United States
14594	Coralliophila	cf. richardi	(P.Fischer		Dredged off Egmont Key, Hillsborough Co., in 200 m	United States
14597	Neoterebra	acrior	Dall	1889	29º14'N., 88º 16' W.; in 100 meters	United States
14650	Neoterebra	limatula	Dall	1889	29º 28'N, 87º 27.30'W; in 173 meters- on boadrR/V "Pelican"	United States
14727	Steironepion	minus	(C.B.Adams	1845	27º 36.65'N, 83º 29.33'W- dredged in 40 m.	United States
14893	Lioglyphostoma	hendersoni	(Bartsch	1934	Dredged in area about 73 miles WSW Anna Maria Is. in 500'	United States
14907	Pazinotus	bowdenensis	Vokes	1970	Dredged WSW of Egmont Key in 240 ft.	United States
15618	Scaphella	gouldiana	(Dall	1887	Dredged in 110 fathoms SE of Palm Beach	United States
15820	Voluta	polypleura aff. form k	Petuch	1981	Banks off San Andrés Is.- crabbed, in fish trap	Colombia
15822	Conus	sahlbergi	da Motta & Harland	1986	Off San Andrés Is.- crabbed, in fish trap	Colombia
15835	Cypraea	nymphae	Jay	1859	7º19'N, 72º25'W-Diego García Is.- on sand/grass, in 6 ft. of water	Diego Garcia
15844	Callistocypraea	broderipii	Sowerby	1832	St. Paul- SCUBA in 110 ft, in coral rubble	Reunion Is.
16089	Voluta	polypleura	Crosse	1876	Roatan,between Camp Bay & Saint Helena- in 8-l0 ft. in sand/grass	Honduras
16130	Parviturboides	species			Isle Dernier (Last Is.), Terrabone Parish- in grunge	United States
16136	Amaea	guineensis	Bouchet & Tillier	1978	Dredged off Cayar, 60 Kms.N. of Dakar	Senegal
16139	Fusimorio =Lox	deshayesii	(Duval	1841	Dredged in 30-40 meters off M'Bour	Senegal
16886	Cancellaria	rosewateri	Petit	1983	29º05.72'N., 88º 22.82'W; in 400-490 m,bottom of mineral nodules	United States
16890	Austrotriton	epitrema	(Tennison-Wood	1879	Dredged off King Is., Tasmania, in 28 ft.	Australia
16910	Rimula	dorriae	Pérez Farfante	1947	29º05.72'N., 88º 22.82'W; in 400-490 m,bottom of mineral nodules	United States
16921	Delectopecten	vitreus	(Gmelin	1791	29º05.72'N., 88º 22.82'W; in 400-490 m,bottom of mineral nodules	United States
16938	Pazinotus	bowdenensis	Vokes	1970	28º37.86'N; 89º 33.25'W; in 38- 104 m.- gray mud	United States
17022	Cornisepta	acuminata	(Watson	1883	29º05.72'N., 88º 22.82'W; in 400-490 m,bottom of mineral nodules	United States
17030	Nodipecten	fragosus	(Conrad	1849	Dredged in 30-50 fathoms by scallop boast off Cape Canaveral	United States
17155	Crenilabium	exile	(Jeffreys	1870	29º05.72'N., 88º 22.82'W; in 400-490 m,bottom of mineral nodules	United States
17156	Cylindriscala	rosenbergi	Garcia	2005	29º05.72'N., 88º 22.82'W; in 400-490 m,bottom of mineral nodules	United States
17309	Adamusium	colbecki	(Smith	1902	Antarctica- in deep water	Antarctica
17336	Siphovasum	latiriforme	(Rehder & Abbott	1951	Dredged in 30-40 fathoms off Contoy Light, Yucatán Peninsula	Mexico
17854	Amaea	tehuanarum	DuShane & McLean	1968	By shrimp nets in 120-240 ft, Bay of Chiriquí	Panamá
18482	Pygmaepterys	germainae	(Vokes & D'Attilio	1980	Puerto Real- in 40 ft.	Puerto Rico
18491	Siratus	vokesorum	García	1999	San Salvador-dredged in 273 m	Bahamas
18503	Siratus	aguayoi	Clench & Farfante	1945	Dredged in 240 m off N. central Cuba	Cuba
18504	Calotrophon	carnicolor	(Clench & Perez Farfante	1945	West of St. Martin, N. of Saba (18º11' N., 63º15' W., in 162 fms	Netherlands Antille
18505	Actinotrophon	actinophorus	(Dall	1889	Gulf of Uraba- dredged in 200 meters by scallop boats	Colombia
18507	Siratus	beaui	Fischer & Bernardi	1857	Dredged in 125 fathoms off Cape San Blas	United States
18508	Paziella	galapagana	Emerson & D'Attilio	1970	Dredged off S. coast of Santa Cruz Galápagos I. in 180-200 meters,	Ecuador
18518	Vokesimurex	tryoni	(Hidalgo	1884	Dredged in 183 m off Key West	United States
18520	Prototyphis	eos paupereques	(Powell	1974	Southern Cave, Poor Knights Ids.- In 50-60 ft.	New Zealand

18521	Timbellus	concavopterus	Kosuge	1980	Mactan Is., Cebu- in tangle nets in 180-200 m.	Philippine Islands
18571	Cirsotrema	hertzae	García	2010	Islas Secas, Golfo de Chiriquí- Dredged in 40-80 ft. in sand	Panamá
19189	Voluta	musica guinaica	Lamarck	1758	Trawled in Amuay Bay, Paraguaná Peninsula- in muddy sand	Venezuela
19190	Voluta	musica	Linnaeus	1758	Trawled off E. coast of Cubagua, S. of Margarita Is.	Venezuela
19229	Favartia	puntagordana	(Weisbord	1962	East side of La Ensenada Beach, S. Isla Chimana Grande,in 3 ft,	Venezuela
20119	Nodipecten	fragosus	(Linnaeus	1758	Off Tarpon Springs, by sponge divers	United States
20152	Euclia	laurettae	Petit & Harasewych	1998	Trawled in 270-360 m, off Isla Montuosa,Golfo de Chiriquí	Panamá
20603	Monstrotyphis	tosaensis	Azuma	1960	Tosa Bay, Shikoku -Dredged in 250 meters	Japan
22130	Turbo	cailletii	Fischer & Bernardi	1856	27º54.895' N 92º 22.594' W, in about 60 meters in calcareous rubble	United States
22165	Vexillum	pulchellum	(Reeve	1844	27º53.418'N 93º18.219W, in coralline rubble in 50-65 meters	United States
22167	Perotrochus	amabilis	(F.M.Bayer	1963	27º53.591'N 91º21.532'W. Alive in 94-155 meters in coralline rubble	United States
22229	Arene	venustula	Aguayo & Rehder	1936	27º 49.24N 92º 53.71W,, in 50-65 m, in rubble.	United States
22246	Zeidora	bigelowi	Farfante	1947	28º 0.561'N 91º 02.205'W, in 58 m.	United States
22272	Purpurellus	macleani	(Emerson & D'Attilio	1969	Pedro González Is., Las Perlas Archipelago	Panamá
22583	Trigonostoma	bullatum	(Sowerby	1832	Trawled in 800-1200 ft, off Punta Mala, Península de Azuero	Panamá
22584	Conus	kohni	Mclean & Nybakken	1979	Trawled in 800-1200 ft, off Punta Mala, Península de Azuero	Panamá
22730	Coralliophila	richardi	(P.Fischer	1882	Dredged in 440 ft. off Key West	United States
22754	Scaphella	dubia dubia	Broderip	1827	Dredged off Key West in 90 fathoms	United States
22756	Scaphella	cuba	Clench	1946	SW of Key West- Dredged in 100 fathoms	United States
23038	Pterotyphis	fimbriatus	(A. Adams	1854	Gobernadora Is., Golfo de Montijo, in 12 ft.	Panamá
23154	Isara	straminea	(A.Adams	1853	28º06.975'N; 90º58.150'W- dredged in 89-92 m in mud.	United States
23164	Conasprella	mindana mindanus f. a	(Dall	1886	28º06.066'N; 91º02.418'W- dredged in 57-65 m in coral rubble	United States
23262	Typhinellus	lamyi	Garrigues & Merle	2014	28º 03.439'N, 92º 26.978'W- in 71- 74 m in coralline rubble	United States
23319	Cycloscala	echinaticosta	(d'Orbigny	1842	27º 59.141'N, 91º 38.832'W- in 91-65 m in coralline rubble with some sp	United States
23321	Favartia	cellulosa	(Conrad	1846	27º 59.141'N, 91º 38.832'W- in 91-65 m in coralline rubble with some sp	United States
23440	Laevinesta	atlantica	Farfante	1947	28º06.066'N; 91º02.418'W- dredged in 57-65 m in coral rubble-	United States
23488	Lioglyphostoma	antillarum	(d'Orbigny	1842	28º20.601'N ; 90º46.926'W- From a core sample in gray mud in 54 m.	United States
23912	Paziella	nuttingi	(Dall	1896	Dredged off ft. Myers in 100 fathoms	United States
24039	Lucinoma	atlantis	(R.A.McLean	1936	27º46.904'N 91º30.286'W, in 546-555 m-hydrocarbon cold seeps	United States
24040	Jorgenia	louisiana	Taylor & Glover	2009	27º46.904'N 91º30.286'W, in 546-555 m-hydrocarbon cold seeps	United States
24041	Cantrainea	macleani	Warén & Bouchet	1993	27º46.904'N 91º30.286'W, in 546-555 m-hydrocarbon cold seeps	United States
24043	Eccliseogyra	formosissima	(Jeffreys	1884	27º46.904'N 91º30.286'W, in 546-555 m-hydrocarbon cold seeps	United States
24054	Mioawateria	malmii	Dall	1889	27º46.904'N 91º30.286'W, in 546-555 m-hydrocarbon cold seeps	United States
24108	Amaea	brunneopicta	(Dall	1908	Off Isla Kativa, W. of Montijo Bay- Dredged in sand in 240 ft.	Panamá
24124	Kanola	meroglypta	McLean & Quinn	1987	27º46.904'N 91º30.286'W, in 546-555 m-hydrocarbon cold seeps	United States
24137	Retimohnia	acadiana	García	2008	27º46.904'N 91º30.286'W, in 546-555 m-hydrocarbon cold seeps	United States
24379	Fenimorea	kathyae	Tippett	1995	28º 6.21'N, 91º 2.23'W- Dredged in 99.3 m.	United States
24383	Cirsotrema	pilsbryi	(McGinty	1940	28º 6.21'N, 91º 2.23'W- Dredged in 99.3 m.	United States
24440	Gymnobela	blakeana	(Dall	1889	27º43.4'N, 91º16.74'W- In 640-680 m, in boxcore	United States
24452	Astyris	verrilli	(Dall	1881	27º43.4'N, 91º16.74'W- In 640-680 m, in boxcore	United States
24458	Emarginula	tuberculosa	Libassi	1859	28º5.24'N, 91º0.4'W-Living inside hard sponge dredged in 66 m.	United States
24459	Anatoma	proxima	(Dall	1927	28º5.63'N, 91º0.67'W- Dredged in 70.8 m; in grunge	United States
24499	Parviturboides	interruptus	(C.B. Adams	1850	28º4.57'N, 90º59.99'W - Dredged in 87.9 m.	United States
24516	Spirolaxis	centrifuga	(Monterosato	1890	28º4.57'N, 90º59.99'W - Dredged in 87.9 m.	United States
24599	Metaxia	espinosai	Rolán & Férnandez-Gárc	1992	28º4.57'N, 90º59.99'W - Dredged in 87.9 m.	United States
24651	Bellaspira	aurantiaca	Fallon	2016	28º4.57'N, 90º59.99'W- Dredged in 87.9 m	United States
24756	Favartia	dipsacus	(Broderip	1833	Playa "El Fraile", Iquique- Diving in 3-5 m.	Chile
24922	Euvola	chazaliei	Dautz	1900	24º35.36'N, 83º24.62'W- Dredged in 64.0-64.1m, in rubble & sand	United States
24929	Boreotrophon	lacunellus	(Dall	1889	27º51.79'N, 84º59.82'W- Dredged in 308-323 m, in fine white silt	United States
24930	Hirtomurex	squamosus	(Bivona	1838	28º38.16'N, 89º33.19'W- Dredged in 60-70 m.-in rubble, with some spon	United States
25047	Coralliophila	aedonia	(Watson	1886	28º17'N, 89º58'W- Wreck of the Tanker "Halo"; In 146.3 m.	United States
25048	Babelomurex	dalli	(Emerson & d'Attilio	1963	28º17'N, 89º58'W- Wreck of the Tanker "Halo"; In 146.3 m.	United States
25090	Seguenzia	cf. formosa f. lineata	Watson	1879	28º39'N, , 88º30'W; in about 1,490.9 m.	United States
25176	Mitrella	profundi	(Dall	1889	Sackett Bank: 28º38.16'N, 89º33.19'W- Dredged in 60-70 m	United States
25216	Niso	hendersoni	Bartsch	1953	Sackett Bank: 28º38.16'N, 89º33.19'W- Dredged in 60-70 m	United States
25218	Calliostoma	barbouri	Clench & Aguayo	1946	Sackett Bank: 28º38.16'N, 89º33.19'W- Dredged in 60-70 m	United States
25228	Astyris	diaphana	Verrill	1882	27º51.79'N, 84º59.82'W- Dredged in 308- 323 m.	United States
25279	Haliris	fischeriana	(Dall	1881	27º51.79'N, 84º59.82'W- Dredged in 308- 323 m.	United States
25299	Anatoma	crispata	(Fleming	1828	27º51.79'N, 84º59.82'W- Dredged in 308- 323 m.	United States
25304	Corinnaeturris	species A			24º35.29'N, 83º41.86'W- Dredged in 496-509 m.	United States
25310	Seguenzia	hapala	Woodring	1928	24º35.29'N, 83º41.86'W- Dredged in 496-509 m.	United States
25320	Calliostoma	cf hirtum	Quinn	1992	24º35.29'N, 83º41.86'W- Dredged in 496-509 m.	United States
25408	Opalia	eolis	Clench & turner	1950	24º43.47'N, 83º13.58'N- Dredged in 61.4-62.1 m.	United States
25455	Sigatica	carolinensis	(Dall	1889	24º43.33'N, 83º13.48'W- Dredged in 61-61.2 m	United States
25457	Vexillum	styria	Dall	1889	27º48.62'N, 84º45.80'W- Dredged in 205- 210.5 m.	United States
25486	Antillipecten Br	antillarum	Recluz	1853	24º35.81'N, 83º28.49'W- Dredged in 61.2-61.3 m.	United States

25498	Conus	attenuatus	Reeve	1844	24º49.06'N, 83º42.80'W- Dredged in 80.7- 81.5m	United States
25514	Hipponix	benthophila	(Dall	1889	24º41.97'N, 83º12.64'W- Dredged in 60- 61 m	United States
25559	Conus	ritae	Petuch	1995	9º22.027'N, 82º14.336'W, Colón Is., Bocas del Toro	Panamá
25656	Conus	glenni	Petuch	2E+07	Bastimento Norte, Bocas del Toro - 9º21'05.2'N, 82º15'34'W	Panamá
25759	Bursa	ranelloides	Dautz & Fisc	1906	24º34.44'N, 83º38.34'W- Dredged in 256-271 m.	United States
25781	Calliostoma	orion	Dall	1889	22º16.07'N, 90º42.89'W- Dreged in 56-57 m, in rubble,	Mexico
25784	Falsuszafrona	yucatecana	García	2023	22º07.22'N, 91º26.43'W- in 45-50 m, in rubble- Gulf of Campeche	Mexico
25786	Oliva	circinata	Marrat	1871	2216.57'N, 90º42.82'W- Dreged in 61-62 m, in sand- Gulf of Campeche	Mexico
25791	Conasprella	rainesae	(Petuch	1987	20º51.49'N, 92º21.44'W- Dredged in 63-65 m, in mud- Campeche	Mexico
25800	Engina	itzamnai	(Watters	2009	20º45.59'N, 92º12.63'W- Dreged in 52- 53 m, in sediment- Campeche	Mexico
25804	Conus	patae	Abbott	1971	22º10.80'N, 91º09.00'W- Dredged in 46-48 m, in rubble- Campeche	Mexico
25811	Turritella	lyonsi	García	2006	20º46.97'N, 91º55.86'W- Dredged in 28-48 m, in mud- Campeche	Mexico
25813	Conasprella	sauros	(García	2006	20º46.97'N, 91º55.86'W- Dredged in 28-48 m, in mud- Campeche	Mexico
25814	Haminoea	solitaria	Say	1822	20º46.97'N, 91º55.86'W- Dredged in 28-48 m, in mud- Campeche	Mexico
25817	Favartia	taylorae	(Petuch	1987	20º51.16'N, 92º26.28'W- Dregded in 93-94 m, iCampeche	Mexico
25819	Monostiolum	harryleei	García	2006	22º16.08'N, 90º42.89'W- Dredged in 54-56 m, in rubble with sponges	Mexico
25830	Conus	patae	Abbott	1971	22º07.98'N, 91º23.75'W- Dredged in 51-48 m, in rubble- Campeche	Mexico
25831	Calotrophon	hystrix	García	2006	22º22.82'N, 90º34.50'W- Dredged in 73-77 m, in hard bottom	Mexico
25835	Fenimorea	sunderlandi	(Petuch	1987	27º58.34'N, 92º22.42'W- Dredged in 68-86 m, in small-size rubble	United States
25844	Vokesimurex	anniae	(M.Smith	1940	22º14.09'N, 91º06.35'W- Dreged in 70 m in hard bottom	Mexico
25846	Calliostoma	orion	Dall	1889	22º16.08'N, 90º42.89'W- in 54-56 m, in rubble with sponges	Mexico
25885	Lucapina	aegis	(Reeve	1850	27º55.72'N, 92º22.58'W- Dreged in 55-150 m, in small-size rubble	United States
25889	Subcancilla	cf. candida	(Reeve	1844	22º46.25'N, 9016.52'W- Dredged in 350 m, in mud	Mexico
25982	Aristofusus	couei	(Petit de la Saussaye	1853	22º16.57'N, 90º42.82'W- Dredged in 61-62 m- LIVE	Mexico
26007	Prunum	rostratum	(Redfield	1870	22º20.00'N, 90º49.43'W- Dredged in 84-89 m	Mexico
26022	Vokesimurex	cabritii	(Bernardi	1859	22º14.28'N, 91º05.98'W- Dredged in 70-68 m, in coarse sand-LIVE	Mexico
26023	Lucapinella	limatula	(Reeve	1850	22º14.28'N, 91º05.98'W- Dredged in 70-68 m, in coarse sand	Mexico
26053	Niveria	harriettae	Fehse & Grego	2010	22º15.15'N, 90º43.29'W- Dredged in 52-53 m, in rubble	Mexico
26084	Laevichlamys	multisquamata	(Dunker	1864	21º06.60'N, 92º08/66'W-- Dredged in 50-45 m- LIVE	Mexico
26098	Busycoarctum	coarctatum	(Sowerby.	1825	22º16.57'N, 90º42.82'W- Dredged in 61-62 m- LIVE	Mexico
26160	Antillipecten Br	antillarum	Recluz	1853	22º10.38'N, 91º09.69'W- Dredged in 49-47 m	Mexico
26207	Epitonium	pigrum	García	2011	20º51.49'N, 92º21.44'W- Dredged in 63-65 m, in mud- Campeche	Mexico
26217	Papyridea	lata	(Born	1778	21º51.32'N, 92º03.68'W- Dredged in 66-68 m, in mud/shells	Mexico
26218	Splendrillia	grandis	Fallon	2016	21º51.32'N, 92º03.68'W- Dreged in 66-68 m, in mud/shells	Mexico
26222	Vexillum	hendersoni	(Dall	1927	28º03.71'N, 92º27.77'W- Dreged in 59-62 m, in midium-size	United States
26225	Cryptopecten	phrygium	(Dall	1886	22º16.21'N, 91º29.14'W- Dredged in 111- 116 m, in mud	Mexico
26244	Volvarina	maya	Espinosa & Ortea	1998	22º11.96'N, 91º15.96'W- Dreged in 89-92 m, in mud	Mexico
26252	Dentimargo	hennequini	Cossignany	2004	22º15.12'N, 90º41.63'W- Dredged in 51-56 m, in rubble and sediment	Mexico
26253	Pseudopusula	antillarum	Schilder	1922	22º15.12'N, 90º41.63'W- Dredged in 51-56 m, in rubble and sediment-LIV	Mexico
26267	Typhinellus	labiatus	(Cristofori & Jan	1832	20º52.40'N, 92º24.83'W- Dredged in 77-81 m in mud, Campeche	Mexico
26283	Opalia	morchiana	(Dall	1889	20º51.49'N, 92º21.44'W- Dredged in 63-65 m, in mud- Campeche	Mexico
26285	Opalia	pumilio	(Mørch	1874	20°00.35'N ; 92°26.10'W, 73-77m, very muddy	Mexico
26291	Opalia	aurifila	(Dall	1889	20º51.16'N, 92º26.28'W- in 93-94 m, in granular sediment	Mexico
26309	Vexillum	garciai	Salisbury & Wolff	2009	20º51.49'N, 92º21.44'W- Dredged in 63-65 m, in mud	Mexico
26310	Simnia	bartschi	(Cate	1973	20°00.35'N ; 92°26.10'W, 73-77m, very muddy	Mexico
26311	Cosmioconcha	geigeri	García	2006	22º16.28'N, 91º30.42'W- Dredged in 108- 107 m., in mud	Mexico
26345	Dentimargo	gibbus	García	2006	22º16.28'N, 91º30.42'W- Dredged in 108- 107 m., in mud	Mexico
26385	Jaspidella	miris	Olsson	1956	22º15.12'N, 90º41.63'W- in 51-56 m, in rubble and sediment	Mexico
26426	Calliostoma	euglyptum	(A.Adams	1854	22º10.80'N, 91º09.00'W- Dredged in 46-48 m, in rubble-	Mexico
26434	Calliostoma	fascinans	Schwengel & McGinty	1992	20°00.35'N ; 92°26.10'W, 73-77m, very muddy	Mexico
26442	Calliostoma	dentatum	Quinn	1992	20º51.49'N, 92º21.44'W- Dredged in 63-65 m, in mud-	Mexico
26503	Rimula	frenulata	(Dall	1889	20º51.16'N, 92º26.28'W- Dregded in 93-94 m, in granular sediment-	Mexico
26507	Lioglyphostoma	aguadillanum	(Dall & Simpson	1901	20°00.35'N ; 92°26.10'W, 73-77m, very muddy	Mexico
26509	Miraclathurella	herminea	Bartsch	1934	20º51.49'N, 92º21.44'W- Dredged in 63-65 m, in mud.	Mexico
26513	Glyphostoma	pilsbryi	Schwengel	1940	20º51.16'N, 92º26.28'W- in 93-94 m, in granular sediment	Mexico
26526	Cerodrillia	occidua	Fallon	2016	20º51.49'N, 92º21.44'W- Dredged in 63-65 m, in mud	Mexico
26570	Eubela	mcgintyi	Schwengel	1943	20º52.40'N, 92º24.83'W- Dredged in 77-81 m in mud,	Mexico
26633	Darrylia	clendenini	García	2008	20º51.49'N, 92º21.44'W- Dredged in 63-65 m, in mud-	Mexico
26636	Leptadrillia	campechensis	Fallon	2016	20º51.49'N, 92º21.44'W- Dredged in 63-65 m, in mud; common	Mexico
26772	Spinosipella	agnes	Simone & Costa	2008	22º47.17'N, 90º15.97'W- Dredged in 350 m, in sediment	Mexico
26809	Poromya	rostrata	Rehder	1943	20º51.16'N, 92º26.28'W- Dregded in 93-94 m, in granular sediment	Mexico
26819	Calliotropis	tiara	(Watson	1879	22º47.17'N, 90º15.97'W- Dredged in 350 m, in sediment	Mexico
27385	Ctenoides	sanctipauli	Stuardo	1982	22º46.25'N, 9016.52'W- Dredged in 350 m, in mud	Mexico
27388	Calliotropis	globosa	Quinn	1991	28º44.43'N, 87º30.53'W, in 1773-1699 m, in mud- Benthic Skimmer	United States
27390	Leucosyrinx	verrillii	(Dall	1881	28º44.43'N, 87º30.53'W, in 1773-1699 m, in mud- Benthic Skimmer	United States
27392	Leucosyrinx	species B	(Dall	1881	28º44.43'N, 87º30.53'W, in 1773-1699 m, in mud- Benthic Skimmer	United States

27396	Leucosyrinx	verrilli	(Dall	1881	28º44.43'N, 87º30.53'W, in 1773-1699 m, in mud- Benthic Skimmer	United States
27397	Gaillea	canetae	(Clench & Aguayo	1944	28º06.52'N , 89º36.99'W- 850- 610 m; in mud. Benthic Skimmer ,	United States
27398	Lucinoma	luteophila	Taylor & Glover	2009	28º06.52'N , 89º36.99'W- 850- 610 m; in mud. Benthic Skimmer ,	United States
27414	Globidrillia	smirna	(Dall	1881	28º06.52'N , 89º36.99'W- 850- 610 m; in mud. Benthic Skimmer ,	United States
27415	Gregorioiscala	pachya	(Locard	1897	28º06.52'N , 89º36.99'W- 850- 610 m; in mud. Benthic Skimmer ,	United States
27416	Cetoconcha	margarita	(Dall	1886	27º35.22'N to 27º33.13'N, 91º52.31'W to 91º47.55'W, in 600-960 m, in h	United States
27569	Cyphoma	intermedium	Sowerby I	1828	28º48.09'N,89º22.89'W- Dredged in 86 m, in mud.	United States
27588	Calliostoma	pulchrum	C.B.Adams	1850	28º15.754'N, 83º44.996'W- Dredged n 33.9 - 33.9 m.-LIVE	United States
27599	Aristofusus	couei	Petit	1853	29º33.59'N, 87º59.6'W- Dredged in 78-70 m, in shell bottom	United States
27622	Terebra	lindae	Petuch	1987	29º39.71'N, 86º13.01'W- Dredged in 61 - 59 m.	United States
27685	Natica	perlineata	Dall	1889	29º19.62'N, 87º46.19'W- Dredged in 97 - 108 m-LIVE	United States
27750	Opalia	burryi	Clench & turner	1950	28º25.79'N, 84º27.47'W- Dredged in 50 50.9 m, in rubble/ sand.-LIVE	United States
27831	Dermomurex	pauperculus	(C.B.Adams	1850	29º24.43'N, 87º58.63'W- Dredged in 74-72 m, in mud	United States
27859	Gregorioiscala	abbotti	Clench & turner	1952	29º19.62'N, 87º46.19'W- Dredged in 97 - 108 m	United States
27870	Benthomangeli	bandella	(Dall	1881	28º31.26'N, 87º40.64'W- Dredged in 2432-2468 m- Benthic Skimmer	United States
27873	Cosmioconcha	nana	García	2007	29º24.43'N, 87º58.63'W- Dredged in 74-72 m, in mud	United States
27889	Spinosipella	cf. agnes	Simone & Costa	2008	29º19.62'N, 87º46.19'W- Dredged in 97 - 108 m	United States
27896	Cardiomya	ornatissima	(d'Orbigny	1853	29º24.43'N, 87º58.63'W- Dredged in 74-72 m, in mud-LIVE	United States
27904	Volvarina	aldeynzeri	Cossignani	2005	28º06.78ºN; 90º55.58ºW- Dredged in 101-99 m., in mud	United States
27931	Opalia	abbotti	Clench & turner	1952	Off Linton, E. of Colón; 9º40N 79º36'W- Dredged in 62 m.; in mud	Panamá
27968	Polygona	nemata	(Woodring	1928	29º43.08'N, 85º53.16'W- Dredged in 39 -38.7 m-	United States
28081	Calliostoma	benedicti	Dall	1889	Dredged 25 mi SW of Key West, Florida, in 250 m.	United States
28090	Calliostoma	frumari	Garcia	2007	24º14'N, 82º09'; dredged 37 kms. SW of Key West, in 200 m.	United States
28422	Cinclidotyphis	myrae	DuShane	1969	Sayulita, Nayarit State; among rocks at low tide	Mexico
28614	Clathrodrillia	garciai	Fallon	2016	28º05.009'N , 91º11.365'W, in 110-109 m; dredged in mud.	United States
28618	Sconsia	grayi	A. Adams	1855	28º04.675ºN, ; 91º38.545'W, in 100-104 m, in mud	United States
28651	Lithopoma	tectum	(Lightfoot	1786	Ewing Bank- 28º05.548'N, 91º01.040W, in 58 m, in rubble	United States
28667	Calyptogena	ponderosa	Boss	1968	27º35.156'N, 91º08.609'W, in 950- 833 m	United States
28783	Typhlomangeli‹	nivalis	(Lovén	1846	27º35.156'N, 91º08.609'W, in 950- 833 m	United States
28812	Heilprinia	timessa	(Dall	1889	28º03.080'N, 91º58.641'W, in 101-100 m, in mud.	United States
28814	Austrotrophon	catalinensis	(Oldroyd	1927	Trawled off San Pedro, in 50 m.	United States
28956	Variciscala	raricostata	(Sowerby	1844	Trawl fishing,	Madagascar
29641	Cochlespira	laurettamarrae	García	2010	Golfo de Chiriquí, trawled in 300 m;	Panamá
29646	Cirsotrema	skoglundae	García	2010	Islas Secas, Golfo de Chiriquí, 60-120 ft.	Panamá
29891	Cycloscala	montrouzieri	García	2004	Ewa, Oahu- SCUBA in 44'	United States
29972	Propeamussiun	simoni	Ghys	2022	28º25.812'N, 89º24.971'W, in 640-576 m mud/shell hash	United States
29973	Atrina	species aff. seminuda	(Lamarck	1819	28º33.698'N, 89º28.894'W, in 334-250 m.	United States
29974	Neopygnodont‹	cochlear	(Poli	1795	28º33.698'N, 89º28.894'W, in 334-250 m.	United States
29983	Aequipecten	glyptus	(Verrill	1882	28º33.698'N, 89º28.894'W, in 334-250 m.	United States
30001	Epitonium	krebsii	(Mörch	1874	28º06.383'N, 90º55.435'W, in 76.8-75.6 m, in rubble	United States
30008	Glossocardia	agassizii	(Dall	1886	28º06.383'N, 90º55.435'W, in 76.8-75.6 m, in rubble	United States
30010	Pleurotomella	pandionis	(A.E.Verrill	1880	28º47.987'N, 88º11.987'W, in 1350 m, in mud- Benthic Skimmer	United States
30063	Astyris	frumarkernorum	García	2009	Dredged off Dry Tortugas in 80-90 ft.	United States
30104	Morum	dennisoni	(Reeve	1842	28º05.781'N, 91º01.623'W; in rubble, 56-57 m.	United States
30108	Phenacovolva	piragua	(Dall	1889	29º26.366'N, 87º34.311'W, in 76-80 m in hard bottom	United States
30109	Pseudosimnia	vanhyningi	(M.Smith	1940	29º15.821'N, 88º20.255'W- 89 m, in rubble Alabama Bank	United States
30113	Polinices	leptaleus	(Watson	1881	28º23.935'N, 89º22.508'W, in 675-765m, in mud	United States
30116	Kanola	meroglypta	McLean & Quinn	1987	28º23.935'N, 89º22.508'W, in 675-765m, in mud	United States
30123	Vexillum	articulatum	(Reeve	1845	28º05.793'N, 91º01.607'W, in56-57 m, in rubble	United States
30125	Cryptogemma	periscelida	(Dall	1889	29º05.979'N, 88º21.167'W, in 303-415 m, in mud	United States
30156	Globidrillia	sp. aff. smirna	(Dall	1881	28º23.935'N89º22.508'W to 28º20.939'N, 89º20.779'W- 675-765 m, in m	United States
30160	Cosmioconcha	rikae	Monsecour & Monsecou	2006	28º07.188'N; 90º52.812'W to 28º07.867'N; 90º49.896'W; 116-117 m, in r	United States
30163	Cuspidaria	obesa	(Loven	1846	28º07.188'N; 90º52.812'W to 28º07.867'N; 90º49.896'W; 116-117 m, in r	United States
30174	Neoterebra	arcas	Abbott	1954	29º16.245'N, 88º37.233'W to 29ª08.900'N, 88ø39.768'W-82-84 m.	United States
30178	Volutomitra	bairdii	(Dall	1889	28º23.935'N89º22.508'W, in 675-765 m, in mud.	United States
30184	Distorsio	clathrata	(Lamarck	1816	29º20.683'N, 88º27.136', in 62-63 m, in shell hash	United States
30196	Distorsio	mcgintyi	Emerson & Puffer	1952	28º38.017'N, 89º33.422'W, in 62 m, in rubble	United States
30310	Distorsio	perdistorta	Fulton	1938	28º05.781'N;91º01.623'W, in rubble, 56-57 m., LIVE	United States
30381	Gregorioiscala	species			Dredged off southern Fiji in 250- 300 m	Fiji
30382	Murdochella	species	n. sp. ?		Dredged off southern Fiji in 250- 300 m	Fiji
30383	Cylindriscala	humerosa	(Schepman	1909	Dredged off southern Fiji in 250- 300 m	Fiji
30386	Gregorioiscala	burchorum	DuSahne	1988	Dredged off southern Fiji in 250- 300 m	Fiji
30388	Papuliscala	cf. japonica	(Okutani	1964	Dredged off southern Fiji in 250- 300 m	Fiji
30393	Conus	attenuatus	Reeve	1844	Dredged west of Dry Tortugas in 70- 80 m	United States
30414	Bulla	solida	Gmelin	1791	Dredged S. of Dry Tortugas	United States
30422	Hyotissa	mcgintyi	(Harry	1985	28º32.088'N; 89º22.396'W, In 688- 776m	United States

30429	Pazinotus	bowdenensis	Vokes	1970	29º15.848'N; 88º20.146'W- In 87.3- 88.3 m	United States
30433	Gaillea	canetae	(Clench & Aguayo	1944	28º32.088'N; 89º22.396'W, In 688- 776m	United States
30434	Cuspidaria	rostrata	(Spengler	1793	27º59.070'N; 91º04.279'W to 27º58.821'N; 91º05.651.W- 148.1- 157 m.	United States
30440	Ameranna	florida	García	2008	29º15.681'N; 88º20.237'W, in 78- 86 m; inside hole in rock.	United States
30446	Conus	aureonimbosus	Petuch	1987	28º05.826'N, 91º01.555'W to 28º05.912'N, 91º01.849'W- in 56-53 m	United States
30456	Turbo	castanea	Gmelin	1791	28º05.826'N, 91º01.555'W, in 56-53 m	United States
30483	Bursa	rhodostoma thomae	(d'Orbigny	1847	28º05.689'N; 91º01.915'W to 28º05.928 'N; 91º02.142W- In 55-56 m.	United States
30486	Benthomangeli	antonia	(Dall	1881	28º54.535'N; 88º12.078'W, in 1136-1124m	United States
30487	Emarginula	sicula	Gray	1825	28º06.345'N; 91º02.148'W, in 94- 99m	United States
30516	Tritonoharpa	cf. lanceolata	(Menke	1828	Farallon Is.; 09º05'N; 80º01'W- NE of Portobelo	Panamá
30526	Opalia	hotessieriana	(d'Orbigny	1842	José Pobre; 9º36'N, 79º38'W; E. of Porto Belo- dead on reef	Panamá
30632	Neotiara	gausapata	Reeve	1815	Barrington Id., SCUBA in 40 ft.- Ameripagos Expedition	Ecuador
30771	Periploma	orbiculare	(Guppy	1882	28º37.9'N, 89º46'W to 28º36.04'N, 89º42.64'W- in 200 m	United States
30775	Neoterebra	nassula	Dall	1889	27º59.28'N, 91º39.28'W, in 77-79 m, in rubble	United States
30783	Ferrocina	garciai	Taylor & Glover	2013	27º56.835'N, 92º01.464'W to 27º56.594'N, 92º01.049'W- in 58-58 m	United States
30786	Scaphella	kieneri	Clench	1946	28º37.9'N, 89º46'W to 28º36.04'N, 89º42.64'W- in 200 m-LIVE	United States
30791	Clathrodrillia	wolfei	(Tippett	1995	28º44.76'N, 90º14.15'W, in 27-28 m, in rubble/sponge	United States
30793	Aristofusus	excavatus	((Sowerby	1880	28º37.750'N, 89º34.703'W to 28º37.850'N, 89º34.010'W- in 100 - 85m.	United States
30794	Calliostoma	yucatecanum	Dall	1881	28º44.76'N, 90º14.15'W, in 27-28 m, in rubble/sponges	United States
30823	Prunum	frumari	Petuch	2012	24º14'N, 82º09'; dredged 37 kms. SW of Key West, in 200 m.	United States
30827	Oocorys	sulcata	Fischer	1883	27º34.153'N; 89º50.224'W, In 1125-1148 m, in mud	United States
30829	Siratus	beauii	P. Fischer & Bernardi	1857	27º55.879'N; 91º12.839'W, in 203-204 m, in mud	United States
30830	Scaphella	robusta	(Dall	1889	27º55.879'N; 91º12.839'W, in 203-204 m, in mud	United States
30840	Diodora	fluviana	(Dall	1889	27º55.879'N; 91º12.839'W, in 203-204 m, in mud	United States
30841	Calliostoma	oregon	Clench & Turner	1960	27º55.879'N; 91º12.839'W, in 203-204 m, in mud	United States
30847	Scaphella	kieneri	Clench	1946	28º08.534'N; 90º53.517'W, in 100 m, in mud.	United States
30871	Retidrillia	pruina	(Watson	1881	27º34.52'N; 89º19.39'W, in 1750-1816m, in mud	United States
30878	Conasprella	armiger	Crosse	1858	28º08.534'N; 90º53.517'W, iln 100 m, in mud-	United States
30880	Conus	villepinii	Fischer & Bernardi	1857	27º55.879'N; 91º12.839'W, In 203-204 m, in mud.	United States
30882	Conasprella	mcgintyi	(Pilsbry	1955	28º08.534'N; 90º53.517'W, in 100 m, in mud	United States
30884	Conus	harlandi	Petuch	1987	Quitasueño Bank, 14º 15'N, 81º 15'W, in lobster trap in about 150 m.	Colombia
30887	Echinophoria	coronadoi	(Crosse	1867	Off Cabo Camarón, in fish trap in 70- 200 m.'	Honduras
30890	Fasciolaria	cf. tephrina	de Souza	2002	Off Cabo Camarón, in fish trap in about 200 m.'	Honduras
30921	Voluta	polypleura f.	Crosse	1876	In fish trap" in deep water" , NENE Honduas- In fish Honduras	Honduras
30955	Trona	stercoraria	(Linnaeus	1758	Point des Almadies, Dakar	Senegal
31030	Pygmaepterys	cf.lourdesae	Gibson-Smith, J.& W.	1963	Traveled off Cabo de la vela in about 80m	Colombia
31060	Calliostoma	scalenum	Quinn	1992	28º05.041'N, 91º01.648'W, In 70-75 m, in rubble	United States
31069	Neoterebra	mugridgeae	García	1999	28º36.617'N, 90º22.501'W, in 28-30 m, in mud	United States
31071	Vexillum	sykesi	McGinty	1955	28º05.041'N, 91º01.648'W, in 70-75 m, in rubble	United States
31085	Diodora	arcuata	Sowerby II	1862	27º57.919'N, 91º14.501'W to 27º58.670'N, 91º16.150'W; 189 m	United States
31087	Clathrodrillia	garciai	Fallon	2016	27º57.919'N, 91º14.501'W to 27º58.670'N, 91º16.150'W; 189 m	United States
31094	Trigonostoma	tenerum	(Philippi	1848	28º04.634'N, 91º02.014'W, in 85-90 m, in rubble.	United States
31106	Fusilaria	garciai	Snyder	2013	Near border of Honduras and Nicaragua; in fish trap in about 150 m	Honduras
31109	Bursa	natalensis	Coelho & Matthws	1970	Off Cape Gracias a Dios, in fish trap in 70 m.	Honduras
31124	Chicoreus	bullisi	Vokes	1974	NE of Cape Gracias a Dios, in fish trap in 200 m.	Honduras
31141	Siphonochelus	longicornis	(Dall	1888	Off Golfo de Morrosquillo- Trawled in 200 m	Colombia
31154	Paziella	pazi	(Crosse	1869	Off Fort Myers- Dredged in 100 fathoms in sand/ fine rubble	United States
31158	Scaphella	dubia schmitti	Clench	1946	Trawled off Savannha in 100 m	United States
31197	Conasprella	rainesae	McGinty	1953	25º01.354'N; 83º43.559'W, in 82-85 m, in rubble	United States
31199	Latiromitra	niveobabelis	Garcia	2015	25º51.104'N, 84º52.278'W, in 1737 m, in rubble/mud	United States
31200	Bathytoma	viabrunnea	(Dall	1889	24º29.650'N, 83º32.149'W to 24º28.317'N, 83º29.226'W, in 353-355 m	United States
31202	Leucosyrix	subgrundifera	(Dall	1888	24º26.631'N, 83º34.136'W to 24º27.699'N, 83º35.950'W- in 557 m.	United States
31234	Falsuszafrona	tortugana	(García	2015	24º48.929'N, 83º40.609'W, in 65 m, in rubble and crushed shells	United States
31248	Calliostoma	roseolum	Dall	1881	25º01.354'N, 83º43.559'W , in 82-85 m , in rubble	United States
31260	Bartschia	frumari	Garcia	2008	25º31.091'N, 84º28.391'W, in 352-361 m, in shells/sand	United States
31264	Glyphostoma	coronaseminale	García	2015	25º31.091'N, 84º28.391'W, in 352-361 m, in shells/sand	United States
31266	Typhlomangeli:	nivalis	(Lóven	1846	25º31.091'N, 84º28.391'W, in 352-361 m, in shells/sand	United States
31267	Lucinoma	filosa	(Stimpson	1851	27º51.79'N, 84º59.82'W- Dredged in 308- 323 m.	United States
31284	Cuspidaria	microrhina	Dall	1886	25º51.104'N, 84º52.278'W, in 1737 m, in rubble/mud	United States
31300	Voluta	polypleura f.	Crosse	1876	In fish trap in 70 m, NE Honduras	Honduras
31301	Voluta	polypleura f.	Crosse	1876	In lobster trap in 50-100 m, NE Honduras	Honduras
31302	Voluta	polypleura	Crosse	1876	Conch diver, in 30-40 ft., NE Honduras	Honduras
31321	Siratus	gundlachi	(Dunker	1883	Off Cape Gracias a Dios-in fish trap in 100 m	Honduras
31323	Heilprinia	coltrorum	Haddorn & Rogers	2000	Off Cabo Camarón, in fish trap in about 200 m.'	Honduras
31366	Voluta	musica	Linnaeus	1758	La Isleta, Margarita Island	Venezuela
31396	Cochlespira	bevdeynzerae	García	2010	Trawled off Cartagena, in 200- 300 m	Colombia

31404	Fasciolaria	delicatissima	García, Lyons & Snyder	2016	North of Cayos de San Andrés- Trawled in 350 m.	Colombia
31425	Murexsul	sunderlandi	(Petuch	1987	Off Playa Forti, in 213 meters, NW Curaçao	Netherlands Antille
31429	Antillophos	smithi	(Watson	1886	Off Playa Forti, in 213 meters, NW Curaçao	Netherlands Antille
31554	Columbella	marrae	García	1999	Cebaco Is., Golfo de Montijo- in beach drift	Panamá
31730	Fusinus	blakensis	Haddorn & Rogers	2000	Blake Plateau, northern bahamas; 400-450 m TOPOTYPE	Bahamas
31892	Voluta	polypleura	Crosse	1876	Trawled off Nicaragua coast	Nicaragua
31893	Voluta	kotorai	Petuch	1981	In Lobster traps; Nicaragua lobster grounds	Nicaragua
31942	Personopsis	grasi	(Bellardi in d'Ancona	1872	27º52.809'N; 91º21.048'W, -in 215 m, in mud	United States
31944	Fusinus	aepynotus	(Dall	1889	27º52.809'N; 91º21.048'W, in 215 m,in mud	United States
31946	Bathyferula	garciai	Stahlschmidt, Lamy Y Fra	2024	27º52.809'N; 91º21.048'W, -in 215 m, in mud	United States
31954	Thylacodes	decussatus	(Gmelin	1791	28º05.212'N, 91º00.628'W- in 68 m; in fine rubble	United States
31960	Lucapina	aegis	Reeve	1850	28º05.423'N, 91º01.253'W, in 60 m in fine rubble	United States
31963	Vokesimurex	sallasi =bellegladensis	(Rehder & Abbott	1951	27º58.183'N, 92º01.881'W-90 m,hard bottom, urchins	United States
31974	Aristofusus	halistreptus	(Dall	1889	28º03.335'N, 92º27.100'W, in 67m; in fine rubble/rocks	United States
31984	Ranularia	rehderi	A.H.Verrill	1950	8º02.508'N, 92º26.872'W, in 64 m in rubble	United States
32000	Bursa	ranelloides	Reeve	1844	27º58.594'N, 91º39.768'W, in 68m; in rubble	United States
32052	Harpa	goodwini	rehder	1992	French Frigate Shoals- Crabbed, in lobster trap in 100 m.	United States
32260	Haliotis	pourtalesii	Dall	1881	27º53.62'N, 91º21.61'W - in 87-95 m-in muddy shale-rock rubble.	United States
32261	Pygmaepterys	richardbinghami	(Petuch	1987	28º04.77'N, 90º59.83'W to 28º04.83'N, 90º59.58'W, in 77m, in rubble	United States
32283	Conus	ermineus	Born	1780	27º53.54'N, 91º21.56'W to 27º53.50'N, 91º21.52'W - in 84-117 m	United States
32285	Phyllonotus	mexicanus	(Petit de la Saussaye	1852	27º53.54'N, 91º21.56'W to 27º53.50'N, 91º21.52'W - in 84-117 m	United States
32286	Isara	ulala	(García	2011	27º53.54'N, 91º21.56'W to 27º53.50'N, 91º21.52'W - in 84-117 m	United States
32289	Conus	stimpsoni	Dall	1902	27º58.19'N, 92º02.03'W to 27º58.31'N, 92º01.81'W- in 86-91m	United States
32299	Entemnotrochu	adansonianus	(Crosee & P. Fischer	1861	27º53.62'N, 91º21.61'W- in 87-95 m-in muddy shale-rock rubble	United States
32305	Siratus	consuela	Verrill	1950	27º53.62'N, 91º21.61'W, in 87-95 m-, in muddy shale	United States
32316	Isara	ulala	(García	2011	28º03.831'N, 9º27.18'W to 28º03.998'N, 92º27.277'W- in 70-68m	United States
32334	Prunum	cassis	(Dall	1889	27º49.34'N, 93º05.92'W- In 195-196 m; in mud	United States
32346	Emarginula	phrixodes	Dall	1927	27º55.14'N, 92º22.71'W to 27º55.09'N, 92º22.08'W- in 168-160 m	United States
32347	Mirachelus	clinocnemus	Quinn	19279	27º55.14'N, 92º22.71'W to 27º55.09'N, 92º22.08'W- in 168-160 m	United States
32359	Cyclopecten	leptaleus	(Verrill	1884	27º55.14'N, 92º22.71'W to 27º55.09'N, 92º22.08'W- in 168-160 m	United States
32362	Eulopia	sagrinata	(Dall	1886	27º55.14'N, 92º22.71'W to 27º55.09'N, 92º22.08'W- in 168-160 m	United States
32384	Leptadrillia	lucaya	Fallon	2016	27º53.65'N, 91º21.48'W to 27º53.3'N, 91º21.2'W- in 219-221 m- in mud	United States
32386	Euchelus	guttarosea	Dall	1889	27º53.65'N, 91º21.48'W to 27º53.3'N, 91º21.2'W- in 219-221 m- in mud	United States
32407	Cosmioconcha	rikae	Monsecour & Monsecou	2006	28º11.80'N, 91º09.852'W to 28º11.865'N, 91)09.781W- in 85-85m	United States
32468	Acesta	oophaga	Järnegren, Schander & Y(	2007	92º25.15'N,27º33.6'W.in 2010 ft.	United States
32532	Epitonium	thorsson	Dushane	1988	Punta Engaño, Mactan I-in tangle nets	Philippine Islands
32638	Voluta	polypleura f.	Crosse	1876	NE Honduras, in lobster trap	Honduras
32775	Scaphella	robusta	(Dall	1889	Dredged off Egmont Key, in 120 fathoms	United States
32777	Scaphella	garciai	Bail	2007	Dredged off Egmont Key, in 120 fathoms	United States
32782	Nodipecten	fragosus	(Conrad	1849	Trawled S. of Pensacola in 80 m.	United States
32784	Siratus	beauii	P. Fischer & Bernardi	1758	Dredged off Pensacola in ll0 fms.	United States
32789	Calliostoma	sayanum	Dall	1889	Tr. in l00 fms. off Cape Catoche, Yucatán	Mexico
32791	Calliostoma	hendersoni	Dall	1927	SW of Key West, in 100 fathoms	United States
32792	Calliostoma	bigelowi	Clench & Aguayo	1938	Dredged S. of Mobile, in 100-110 fathoms	United States
32800	Conus	largillierti	Kiener	1848	Tr. in 40 fms off Contoy Light, Yucatán	Mexico
32807	Prunum	cassis	(Dall	1889	N. Contoy; trawled in 100 m	Mexico
32809	Calliostoma	marionae	Dall	1906	S 30.7 mmW. of Egmont Key, St. Petersburg; dredged in 30-40 fathoms	United States
32810	Conasprella	sauros	(García	2006	Tr. in 25-30 fms. off Cabo Catoche, Yucatán	Mexico
32823	Amaea	retifera	(Dall	1889	Tr. in 40 fms. off Cabo Catoche, Yucatán	Mexico
32827	Scaphella	dubia dubia	Broderip	1827	Dredged off Egmont Key, in 110 fathoms	United States
32830	Oocorys	clericus	Quinn	1980	Tr. SE of South Pass, Mississippi River Delta in 200-250 m	United States
32846	Trigonostoma	tenerum	(Philippi	1848	Off Egmont Key, in 90 m.	United States
32847	Eucorys	bartschi	Rehder	1943	Off Contoy Light, in 200-365 m	Mexico
32854	Sconsia	nephele	F.M.Bayer	1971	Rapid Point, Union Island, SCUBA in 40-50 ft.- at night	Saint Vincent
33024	Semicassis	labiatum	(Perry	1811	False Bay, Capetown- SCUBA in 40 ft.	South Africa
33024	Nodipecten	fragosus	(Conrad	1849	In fish trap 0ff Key West	United States
33030	Voluta	polypleyra	Crosse	1876	Tr. in Honduras shrimping grounds	Honduras
33039	Scaphella	dubia schmitti	(Bartsch	1931	Dredged 16 mi SW of Key West, in 682 ft-smaller, live	United States
33042	Scaphella	robusta	(Dall	1889	Dredged SW of Key West in 757 ft	United States
33043	Scaphella	dubia georgiana	Clench	1946	Dredged SW of Key West in 757 ft	United States
33044	Scaphella	gouldiana	Dall	1887	Dredged SW of Key West in 125 fathoms-	United States
33049	Scaphella	gouldiana	Dall	1887	Trawled SE of St. Augustine by shimp boat in 1100 ft.	United States
33053	Fasciolaria	bullisi	Lyons	1972	NW of Dry Tortugas in lobster trap in 250 ft.	United States
33078	Cryptopecten	phrygium	(Dall	1886	Off Sand Key, in 300 ft-by shrimp boat "Michael James"	United States
33082	Dolicholatirus	pauli	(McGinty	1953	26 Km SW of Key West, in 250 m, in very hard bottom	United States
33088	Dilemma	frumarkernorum	Leal	2000	Dredged SW of Key West in 125 fathoms	United States

33094	Niso	tricolor	Dall	1889	16 mi SW of Key West-dredged in 580 ft.	United States
33097	Cochlspira	cavalier	García	2010	S. of Dry Tortugas in 180-200 fathoms- in sandy mud	United States
33098	Discotechtonica	discus	(Philippi	1844	Dredged 26 Km SW of Key West, in 250 m, in very hard bottom	United States
33099	Adelphotectoni	uruguaya	(Carcelles	1953	Dredged 26 Km SW of Key West, in 250 m, in very hard bottom	United States
33107	Sthenorytis	pernobilis	(Fischer.& Bernardi	1857	Dredged 26 Km SW of Key West, in 250 m, in very hard bottom	United States
33109	Calliostoma	marionae	Dall1963)	1906	35 mi SW of Key West- in fish tap in 250 ft	United States
33238	Timbellus	havanensis	Vokes	1970	Dredged 16 mi SW of Key West, in 598 ft.	United States
33245	Cirsotrema	cf. dalli	Rehder	1945	7 mi off Bahia Honda, Florida Keys, shell dredge in 220 ft.	United States
33249	Dolichupis	leei	Fehse & Grego	2010	Off Boca Chica, Florida Keys, in 240 ft.-with shelldredge	United States
33256	Punctiscala	watsoni	(de Boury	1911	SW of Key West, in 757 ft.	United States
33315	Epitonium	celesti	(Aradas	1854	SW of Sand Key, off Ket West, in 400 ft.	United States

REFERENCES

García, E. F. 1988. Observations on the *Voluta demarcoi* complex and the genus *Voluta* in the western Caribbean. *Louisiana Malacological Society. Occasional Papers* No. I, 1-7.

García, E. F. 1989. Two rare volutes from Yucatán. *American Conchologist* 17 (3):3-4.

García, E. F. 1993. The genus *Voluta* in the Caribbean. *American Conchologist* 21(2):12-17.

García, E. F. 1997. A rare pair and other jewels from the northern Gulf of Mexico. *American Conchologist* 25(1):8-9.

García, E. F. 1999a. New molluscan records for the northwestern Gulf of Mexico. *American Conchologist* 27(2):27-28.

García, E. F. 1999b. Three new gastropod species from the New World. *Apex* 14(3-4):59-65.

García, E. F. 2000. Surprising new molluscan records from Louisiana and the northwestern Gulf of Mexico. *American Conchologist* 28(3):5-6.

García, E. F. 2000b. A rare pair. *American Conchologist* 28(4):7.

García, E. F. 2001. A remarkable deep-water *Cochlespira* from Golfo de Chiriquí, southwestern Panamá. *The Festivus* 33(3):23-24.

García, E. F. 2002a. More discoveries from a collecting expedition off the Louisiana coast. *American Conchologist* 30(1):6-8.

García, E. F. 2002b. Unexpected molluscan finds from the hydrocarbon vents off the Louisiana coast. *American Conchologist* 30(4):28-30.

García, E. F. 2004. On the genus *Cycloscala* Dall, 1889 (Gastropoda: Epitoniidae) in the Indo-Pacific, with comments on the type species, new records of known species, and the description of three new species. *Novapex* 5(2-3):57-68.

García, E. F. 2005. Six new deep-water molluscan species (Gastropoda: Epitoniidae, Conoidea) from the Gulf of Mexico. *Novapex* 6(4):79-87.

García, E. F. 2006a. *Conus sauros*, a new *Conus* species (Gastropoda: Conidae) from the Gulf of Mexico. *Novapex* 7(2-3):71-76.

García, E. F. 2006b. Six new species of mollusks (Gastropoda: Cerithioidea, Buccinoidea, Muricoidea) from Bahía de Campeche, southwestern Gulf of Mexico. *Novapex* 7(4):77-89.

García, E. F. 2007a. Report on mollusks collected in a dredging expedition to Bahía de Campeche, southwestern Gulf of Mexico. *American Conchologist* 35(2)4-11.

García, E. F. 2007b. Results of deep-water dredging in the Gulf of Mexico using the "Benthic Skimmer", and report on several geographic extensions, including two species not previously reported in the western Atlantic. *The Festivus* 39(2):13- 18.

García, E. F. 2007c. A new species of *Cosmioconcha* (Gastropoda: Columbellidae) from the northern Gulf of Mexico. *Novapex* 8(2):43-46.

García, E. F. 2007d. Three new deep-water species of mollusks (Gastropoda: Calliostomatidae, Cystiscidae) from the southeastern Gulf of Mexico. *Novapex* 8(3-4):127-133.

García, E. F. 2008a. Eight new molluscan species (Gastropoda: Turridae) from the western Atlantic, with the description of two new genera. *Novapex* 9(1):1-15.

García, E. F. 2008b. On two *Opalia*-like epitoniids (Gastropoda: Epitoniidae) from the Gulf of Mexico. *The Festivus* 40(7):83-86.

García, E. F. 2008c. Frustrations and extensions IV: On two *Cyphoma* species from the Gulf of Mexico and six geographical extensions. *American Conchologist* 36(1):4-8.

García, E. F. 2008d. An extension of the genus *Spinosipella* (Bivalvia: Verticordidae) in the Gulf of Mexico. *American Conchologist* 36(3):8-9.

García, E. F. 2008e. Four new buccinid species (Gastropoda: Buccinidae) from the western Atlantic. *Novapex* 9(4):141-148.

García, E. F. 2008f. Molluscan findings from a recent dredging expedition off the Louisiana coast. *American Conchologist* 36(4):4-9.

García, E. F. 2009. A new *Astyris* species (Gastropoda: Columbellidae) from the Gulf of Mexico, with notes on the genus. *Novapex* 10(1):5-8.

García, E. F. 2010a. The genus *Cirsotrema* (Gastropoda: Epitoniidae) in the Panamic Province, with the description of two new species. *Novapex* 11(4):93-98.

García, E. F. 2010b. Description of four new species of *Cochlespira* (Gastropoda: Turridae) from the New World. *Novapex* 11(4):107-113.

García, E. F. 2011a. A new species of *Mitra* (*Fusimitra*) (Gastropoda: Mitridae) from the. northwestern Gulf of Mexico. *Novapex* 12(3-4):57-62.

García, E. F. 2011b.Two new species of *Epitonium* (Gastropoda: Epitoniidae) from the western Atlantic. *Novapex* 12(3-4):99-107.

García, E. F. 2011c. Noteworthy mollusks of the Gulf of Mexico, including record sizes and geographical extensions. *American Conchologist* 39(3):4-9.

García, E. F. 2011d. On *Vexillum (Pusia) articulatum* (Reeve, 1845) and *V. (P.) trophonium* (Dall, 1889). *American Conchologist* 39(3):10-11.

García, E. F. 2012a. Noteworthy offshore mollusks from the north- central Gulf of Mexico, including geographical extensions and a generic reassignment. *American Conchologist* 40(1):34- 37.

García, E. F 2012b. *Terebra limatula* Dall, 1889 and *T. acrior* Dall, 1889 (Gastropoda: Terebridae); two problematic taxa from the western Atlantic. *Zootaxa* 3328:66-68.

García, E. F 2012c. Report on a dredging expedition off the Louisiana coast including geographical extension and new record size. *American Conchologist* 40(4):7-10.

García, E. F 2013a. Report on a dredging expedition off the Louisiana coast including geographical extension and new record size. II. *American Conchologist* 41(1):4-8.

García, E. F 2013b. *Turritella lyonsi* found off the Louisiana coast. *American Conchologist* 41(4):8-9.

García, E. F 2015a. A new species of *Zafrona* (Gastropoda: Columbellidae) from the Gulf of Mexico. *Novapex* 16(2):49-52.

García, E. F 2015b. A new species of *Latiromitra* (Gastropoda: Ptychatractidae) from the Gulf of Mexico. *The Nautilus* 129(2):90-93.

García, E. F 2015c. A new species of *Glyphostoma* (Gastropoda: Clathurellidae) from the Gulf of Mexico. *The Nautilus* 129(3):136-139.

García, E. F 2016a. The genera *Miraclathurella* Woodring., 1928 (Gastropoda: Pseudomelatomidae) and *Darrylia* García, 2008 (Gastropoda: ?Horaiclavidae), with two proposed new combinations for *Darrylia*. *The Nautilus* 130(2):79-81.

García, E. F 2018. Report on a dredging expedition off the Louisiana coast including surprising geographical extensions. *American Conchologist* 46(3):8-12.

García, E. F 2019. Report on a dredging expedition off the Louisiana coast including surprising geographical extensions II. *American Conchologist* 47(4):4-11.

García, E. F 2021a. Report on mollusks collected in a single dredge haul in the Gulf of Mexico. What does it say about future molluscan discoveries? *American Conchologist* 49(2):16-21.

García, E. F 2021b. Species of the family Borsoniidae dredged by the R/V *Pelican* in the Gulf of Mexico. *American Conchologist* 49(4):8-12.

García, E. F 2022a. On two species of *Amaea* (Gastropoda: Epitoniidae) from the western Atlantic Ocean. *American Conchologist* 50(2):8-11.

García, E. F 2022b. Species of Buccinidae collected by the R/V *Pelican* in the Gulf of Mexico. *American Conchologist* 50(1):20-23.

García, E. F 2022c. Species in the families Cystiscidae, Marginellidae, and Granulinidae dredged by the R/V *Pelican* in the Gulf of Mexico. *American Conchologist* 50(4):20.

García, E. F 2023a. Species of the families Costellariidae and Mitridae dredged in the Gulf of Mexico by the R/V *Pelican* in the Gulf of Mexico. *The Festivus* 55(1):65-70.

García, E. F 2023b. Species of Muricidae (Gastropoda: Muricoidea) dredged by the R/V *Pelican,* and the synonymy of two well-known names. *Novapex* 24(1):21-32.

García, E. F 2023c. Species of the families Conidae and Terebridae dredged by the R/V *Pelican* in the Gulf of Mexico. *American Conchologist* 51(1):11-22.

García, E. F 2023d. *Falsuszafrona yucatecana* (Gastropoda: Columbellidae), a new species from the southwestern Gulf of Mexico. *The Nautilus* 137(2):52-54.

García, E. F 2023e. Species of the family Fissurellidae dredged by the R/V *Pelican* in the Gulf of Mexico. *American Conchologist* 51(3):4-8.

García, E. F. 2023f. Species of Fasciolariidae, Dolicholatiridae, Busyconidae and Coumbellidae dredged by the R/V *Pelican* in the Gulf of Mexico. Gloria Maris 62(2): 58-66.

García, E. F. 2023g. Species of the superfamilies Trochoidea (Gastropoda: Areneidae, Calliostomatidae, Colloniidae, Liotiidae, Skeneidae, Solariellidae, Trochidae, Turbinidae) and Seguenzioidea (Gastropoda: Calliotropidae, Cataegidae,Chilodontidae, Cataegidae, Chilodontidae, Seguenzidae) dredged by the R/V *Pelican* in the Gulf of Mexico. *Xenophora Taxonomy* 41:17- 23.

García, E. F. 2024a. A new species of *Corinnaeturris* (Gastropoda: Clathurellidae) from the southeastern Gulf of Mexico. *The Festivus* 56(1):27-30.

García, E. F. 2024b Species of the families Epitoniidae *and* Architectonicidae (Mollusca: Gastropoda) dredged by the R/V *Pelican* in the Gulf of Mexico. *Novapex* 25(1):1-9.

García, E. F. 2024c. A shell collector in a nut shell. *American Conchologist* 52(1):29.

García, E. F. 2024d. Species of the families Drilliidae Olsson, 1964 and Clathurellidae H. Adams & A. Adams, 1858 (Gastropoda: Conoidea), collected by *the* R/V *Pelican* in the Gulf of Mexico. *American Conchologist* 52(3):10-16.

García, E. F. and W. K. Emerson. 1987. A new species of *Scaphella* (Gastropoda: Volutidae) from off Yucatán, Mexico. *Apex. Informations Scientifiques de la Société Belge de Malacologie*11(1):1-5.

García, E. F. and H. G. Lee. 2002. Report on molluscan species found in the offshore waters of Louisiana, including many extensions of known range and un-named species. *American Conchologist* 30(4):10-13.

García, E. F. and H. G. Lee. 2003. Report on molluscan species found in the offshore waters of Louisiana, including many extensions of known range and un-named species. II. *American Conchologist* 31(1):26-29.

García, E. F. and H. G. Lee. 2004. Report on the malacofauna of offshore Louisiana waters, including many range extensions and un-named species. III. *American Conchologist* 32(3):21-24.

García, E. F., W.G. Lyons & M. A. Snyder. 2016.The western Caribbean complex of *Fasciolaria tephrina* de Souza, 2002 (Gastropoda: Fasciolariidae), with the description of a new species. *The Nautilus* 130(3):127-131.

APPENDIX

Curriculum Vitae

ADDRESS: Home: 115 Oak Crest Dr., Lafayette, LA 70503

Academic: University of Louisiana at Lafayette, Dept. of Biology, P. O. Box 42451, Lafayette, Louisiana 70504, U.S.A.

PERSONAL: Born 22 June, 1939 in Havana, Cuba

EDUCATION:

La Progresiva High School, Cárdenas, Cuba - Diploma, 1956
Louisiana State University (Chemical Engineering Curriculum), 1956-60
University of Southwestern Louisiana, 1960-63; B.A., May 1963
Tulane University, 1963-65; M.A., May 1969
Louisiana State University, 1970-73; Ph.D., August 1973.

TRAVEL:

Most of the United States, most of Mexico, eastern and western Canada, Belize, Guatemala, Costa Rica, Nicaragua, Honduras, Panama, Puerto Rico, Cuba, Haiti, Dominican Republic, St. Martin, Martinique, Guadeloupe, Dominica, the Netherlands, St. Lucia, St. Vincent, Grenada, Colombia, Venezuela, Perú, Bolivia, Ecuador, Brazil, Chile, Juan Fernández Archipelago (Chile), Argentina, Paraguay, Australia, New Zealand, Philippines, Fiji, Tonga, Samoa, Palau Islands, Guam, Rota Is., Tinian Is., Solomon Islands, Vanuatu, Thailand, Cambodia, Malaysia, Indonesia, Majuro, Mili Atoll, Pohnpei, Truk Atoll, Tunisia, French Polynesia (Society Islands and Tuamotu Archipelago), Borneo, Burma, Madagascar, Cape Verde Archipelago, Senegal, Ivory Coast, Ghana, France, Spain, Portugal, Croatia, Montenegro, Albania, Greece, China, Egypt, Jordan, South Africa, Zambia, Swaziland, Zimbabwe, Botswana, India, Nepal, Vietnam, Antarctica, Burma (Myanmar), Germany, Switzerland, Belgium, Tunisia, Russia, Morocco.

TEACHING EXPERIENCE:

Teaching Assistantship, Tulane University, 1963-65
Ursuline Academy, New Orleans, 1964-65
NDEA Summer Institute, Lexington, Virginia, 1965
University of Southwestern Louisiana, 1965-70
Teaching Assistantship, Louisiana State University, 1970-71
University of Southwestern Louisiana, 1971-1997

PROFESSIONAL MEMBERSHIPS AND HONORS:

Sigma Delta Pi, Spanish Honorary Society
Pi Delta Phi, French Honorary Society
Phi Sigma Iota, Foreign Language Honorary Society
Phi Kappa Phi, Honorary Society
American Association of Teachers of Spanish and Portuguese

South Central Modern Language Association
Sabbatical leave for research, Summer 1980
Distinguished Professor Award, 1982
Order of the Discoverers, award given by Sigma Delta Pi, National Spanish Honor Society, 1985
Conchologists of America's 2001 Neptunea Award for service to the science of Malacology
Invitation by the Muséum National d'Histoire Naturelle, Paris, to do research and publish on mollusks in their collection of the family Epitoniidae
Editorial Board, *American Conchologist*
Editorial Board *The Festivus*

HONORARY EPONYMS

Conus garciai daMotta, 1982
Vokesimurex gaciai (Petuch, 1986)
Cerithioclava garciai Houbrick, 1986
Voluta garciai (Petuch, 1987)
Opalia garciai Kilburn, 1994
Sinezona garciai Geiger, 2006
Scaphella garciai Bail, 2007
Vexillum garciai Salisbury & Wolff, 2009
Stocisia garciai Rolán, Fernández-Garxés & Lee, 2009
Anatoma emilioi Geiger, 2011
Haplocochlias garciai Rubio, Fernández-Garcés & Rolán, 2013
Ferrocina garciai Taylor & Glover, 2013
Fusilaria garciai Snyder, 2013
Pusula garciai Fehse & Grego, 2014
Clathrodrillia garciai Fallon, 2016
Garcíavoluta Petuch & Berschauer, 2023, new subgenus
Bathyferula garciai Stahlschmidt & Fraussen, 2024

NEW MOLLUSCAN SPECIES DESCRIBED:

Amaea apexroseus García, 2003
Amaea boucheti García, 2003
Amaea diluta García, 2003
Amaea elegantula García, 2003
Amaea lennyi García, 2003
Boreoscala ponderosa García, 2003
Cirsotrema excelsum García, 2003
Cirsotrema richeri García, 2003
Cirsotrema herosae García 2003
Claviscala pellisanserina García, 2003
Claviscala vivienneae García, 2003
Cylindriscala paradoxa García, 2003
Epitonium boutetorum García, 2016
Epitonium deschampsi García, 2003
Epitonium juanitae García, 2003
Epitonium kastorae García, 2003
Gregorioiscala nivellei García, 2003

Gyroscala mikeleei García, 2003
Periapta weilei García, 2003
Epitonium maestrati García, 2003
Nassarius harryleei García, 2001
Conus saurus García, 2006
Cirsotrema ernestoilaoi García, 2001
Cycloscala aldeynzeri García, 2001
Epitonium bevdeynzerae García, 2001
Colubella marrae García, 1999
Siratus vokesorum García, 1999
Terebra mugridgeae García, 1999
Latirus aldeynzeri García, 2001
Cirsotrema bennettorum García, 2000
Gregorioiscala barazeri García, 2004
Gregorioiscala cosnieri García, 2004
Gregorioiscala levismaculosa García, 2004
Gregorioiscala xanthotaenia García, 2004
Opalia dushaneae García, 2004
Opalia felderi García 2004
Opalia longisssima García, 2004
Opalia megalodon García, 2004
Opalia neocaledonica García, 2004
Oplaia thorsenae García, 2004
Opalia turnerae García, 2004
Opalia velumnuptialis García, 2004
Opalia wareni García, 2004
Cycloscala armata Garcia, 2004
Cycloscala montrouzieri García, 2004
Cycloscala sardelae García, 2004
Cylindriscala rosenbergi García, 2005
Opaliopsis rabalaisi García, 2005
Gymnobela felderi García 2005
Gymnobela fredericqae García, 2005
Gymnobela petiti García, 2005
Anticlinura atlantica García, 2005
Turritella lyonsi García, 2006
Monostiolum harryleei García 2006
Cosmioconcha geigeri García, 2006
Calotrophon hystrix García, 2006
Dentimargo gibbus García, 2006
Canalispira aurea García, 2006
Scaphella mcginnorum García & Emerson, 1987
Cosmioconcha nana García, 2007
Canalispira kerni García, 2007
Canalispira lipei García, 2007
Calliostoma frumari García, 2007
Darrylia harryleei García, 2008
Euthria vokesi Fraussen & García, 2008
Suturocythara apocrypha García, 2008

Suturocythara redferni García, 2008
Miraclathurella clendenini García, 2008
Acmaturris annaclaireleeae García, 2008
Acmaturris pelicanus García, 2008
Agathotoma exthymata García, 2008
Viridrillia aureofasciata García, 2008
Ameranna florida (García, 2008)
Manaria burkeae García, 2008
Bartschia frumari García, 2008
Retimohnia acadiana García, 2008
Astyris frumarkernorum García, 2009
Cirsotrema skoglundae García, 2010
Cirsotrema hertzae García 2010
Cochlespira leeana García, 2010
Cohclespira bevdeynzerae García, 2010
Cochlespira laurettamarrae García, 2010
Cochlespira cavalier García, 2010
Eccliseogyra brasiliensis García, 2011
Epitonium pigrum García, 2011
Epitonium leali García, 2011
Mitra ulala García, 2011
Zafrona tortugana García a, 2015
Latiromitra niveobabelis García, 2015
Glyphostoma coronaseminale García, 2015
Fasciolaria delicatissima García, Lyons & Snyder, 2016
Bathytoma cygnus García, 2021
Falsuszafrona yucatecana García, 2023
Fenimorea falloni García, 2023
Corinnaeturris humilis García, 2024

NEW GENERA DESCRIBED:

Darrylia García, 2008
Suturocythara García, 2008
Panamitra García & Salisbury, 2024

PROFESSIONAL ACTIVITIES:

Past Foreign Language Campus Representative, South Central Modern Language Association
State Director, Sigma Delta Pi, National Spanish Honorary Society
Secretary, Section on Latin American Literature, South Central Modern Language Association, Houston, Texas 1976
Chairman, Section on Latin American Literature, South Central Modern Language Association, Hot Springs, Arkansas, 1977
Chairman, Section on Latin American Literature, Louisiana Conference on Hispanic Language & Literatures, Baton Rouge, 1982
Editor, Entre Nosotros, National Bulletin of Sigma Delta Pi, Spanish Honor Society, 1980-1995
Member, Editorial Board, American Conchologist, an international quarterly bulletin of the Conchologists of America
Judge for several years of Sigma Delta Pi's literary contests for the Southwest

COURSES TAUGHT:

Beginners, Intermediate and Advanced Spanish, Spanish American Culture and Civilization, Survey of Latin American Literature, Introduction to Literature, Modernismo, Twentieth Century Poetry, Twentieth Century Fiction, Foreign Literature in Translation.

WRITINGS & TALKS (Literature):

"El simbolismo del Curandero-Venado de las Siete-Rozas en Hombresde maíz de Miguel Angel Asturias." South Central Modern Language Association, Fort Worth, Texas, 1972.

"The Loss of Panama and Colombian Poetry of the Early Twentieth Century" (discussant). South Eastern Conference of Latin American Studies, 1974.

"Luis Gonzaga y Gabriel: religiosidad y sensualidad en Al filo del agua." South Central Modern Language Association, New Orleans, 1975.

"El arte de la conversación/El arte de la composición." The Modern Language Journal, XLI (September-October 1977), [book review].

"Luis Andry Reports on the Attakapas, 1773." Attakapas Gazette, XIII (Spring, 1978).

"Hombres de maíz: unidad y sentido a través de sus símbolos mitológicos". Miami: Ediciones Universal, 1978) [**book**].

"Los personajes celestiales en El Señor Presidente." Louisiana Conference on Hispanic Language and Literatures, Baton Rouge, 1979.

"Teaching Foreign Literature as an Interdisciplinary Subject." American Association of Teachers of Spanish and Portuguese, 1980.

"La noción existencial del absurdo en Cien años de soledad." South Central Modern Language Association, Austin, Texas, 1981.

"Macondo: realidad paradójica." Louisiana Conference on Hispanic Language and Literatures, Baton Rouge, 1982.

"La Piojosa Grande, María Tecún e Isaura Terrón: diosas luniterrestres de la mitología maya-quiché." Modern Language Association, New York, December 1983.

"El mundo desrealizado de Cien años de soledad." Symposium on Gabriel García Márquez, Mississippi State University, April 1984.

"La noción existencial del absurdo en Cien años de soledad." INTI, Nos. 16-17 (1983), pp. 125-134.

"El mundo desrealizado en Cien años de soledad." En punto de mira: Gabriel García Márquez. Madrid: Editorial Pliegos, 1985, pp. 115-123.

"Colibrí: peregrinación onírica de Severo Sarduy." Mountain Interstate Foreign Language Conference, October 3-5, 1985.

Quest for the Promised Land. Official Correspondence Relating to the First Acadian Migration to Louisiana, 1764-1769. García, Emilio F., et al. transl. Lafayette: The Center for Louisiana Studies, 1989, 175 pp [**book**].

WRITINGS & TALKS (Malacology):

TALKS:

"The Archaeogastropoda: unifying characteristics." Louisiana Malacological Society, Baton Rouge, November, 1982.

"The genera in the family Volutidae." Louisiana Malacological Society, Baton Rouge, October, 1983.

"The family Ranellidae." Louisiana Malacological Society, Baton Rouge, March, 1984.

"Uncommon and rare molluscs of the Western Caribbean." Louisiana Malacological Society, Baton Rouge,

September, 1985.

"The families Mitridae and Costellariidae: similarities and differences." Louisiana Malacological Society, Baton Rouge, April 1986.

"How to identify bivalves." Louisiana Malacological Society, Baton Rouge, October, 1986.

"The family Strombidae: problematic species." Louisiana Malacological Society, Baton Rouge, February, 1987.

"A Key to the genus *Harpa*." Louisiana Malacological Society, Baton Rouge, September, 1988.

"Molluscan fauna of a south Pacific Atoll." Louisiana Malacological Society, Baton Rouge, October, 1990.

WRITINGS:

Fraussen, Koan & E, F, García. 2008. A new and unique *Euthria* (Gastropoda: Buccinidae) from Somalia. *Gloria Maris* 47(3):48-52.

García, E. F. 1971. In search of *Cypraea mus* Linne, 1758. *Keppel Bay Tidings* (June-July), pp. 5-6.

García, E. F. 1974. A variation of *Conus ranunculus* Hwass. *Of Sea and Shore* 5(3):121.

García, E. F. 1975. Shelling on Saint Martin (Netherlands Antilles). *Of Sea and Shore* (Spring, 1975), pp. 37-38.

García, E. F. 1979. Belize and its molluscan fauna. *Of Sea and Shore* (Fall, 1979), pp. 183-186.

García, E. F. 1980. Trawling for Central American mollusks. *Hawaiian Shell News* 28(2):5.

García, E. F. 1980. Variability of *Voluta musica* L. *Of Sea and Shore* (Fall, 1980), pp. 151-152.

García, E. F. 1986. Shell Collecting in Fiji. *Texas Conchologist* 21(2):49-54.

García, E. F. 1988. Observations on the *Voluta demarcoi* complex and the genus *Voluta* in the western Caribbean. *Louisiana Malacological Society. Occasional Papers* No. I, 1-7.

García, E. F. 1988. A rediscovered *Columbella* from the Caribbean. *Hawaiian Shell News* 36(6):12.

García, E. F. 1989. *Spondylus erinaceus* Reeve, an ignored Caribbean species. *American Conchologist* 7(1):9.

García, E. F. 1989. The Caribbean Province. *The Junonia* (March-April, 1989), pp. 6-7.

García, E. F. 1989. Two rare volutes from Yucatán. *American Conchologist* l7 (3):3-4.

García, E. F. 1989. On *Spondylus gilvus* Reeve. *American Conchologist* l7(3):8-9.

García, E. F. 1990. The Turneffe Islands: A shell collecting experience. *The Junonia* (December, l990), pp. 4-5.

García, E. F. 1990. Palau revisited. *American Conchologist* l8(2):l0-11.

García, E. F. 1990. Some thoughts on species differentiation. *American Conchologist* l8(4):6-7.

García, E. F. 1991. A Frech Polynesian atoll and its molluscan Fauna. *American Conchologist*, 21(1):16-19.

García, E. F. 1991. On Shell Show exhibits. *Suncoast Shorelines* 8:1-2.

García, E. F. 1991. An albino Mitra papalis?. *Hawaiian Shell News* 39(5):8.

García, E. F. 1992. Latin l0l (Formation of scientific names). *American Conchologist* 20(4):l8.

García, E. F. 1993. Latin l02 (Formation of scientific names). *American Conchologist* 21(1):5.

García, E. F. 1993. The genus *Voluta* in the Caribbean. *American Conchologist* 21(2):12-17.

García, E. F. 1994. Shelling in Samoa. *American Conchologist* 22(1):16-18.

García, E. F. 1995. On an island far, far away. *American Conchologist* 23(1):23-25.

García, E. F. 1995. Something for everyone. *American Conchologist* 23(l):28.

García, E. F. 1995. Frustrations & Extensions: Problematic & ignored species, and re-definition of two geographical boundaries. *American Conchologist* 23(4):25-26.

García, E. F. 1996. Frustrations & Extension (Part II). *American Conchologist* 2(1):3-5.

García, E. F. 1996. The rediscovery of *Cyclothyca corrugata* Stearns, l990. *American Conchologist* 24(4):19.

García, E. F. 1997. A rare pair - and other jewels from the northern Gulf of Mexico. *American Conchologist* 25(1):8-9.

García, E. F. 1997. *Terebra achates* Weaver, l960: a geographic extension. *American Conchologist* 25(4)21.

García, E. F. 1997. On *Phalium pilsbryi*, an interesting Panamic species. *American Conchologist* (26(1):28.

García, E. F. 1998. A Panamanian Adventure. *American Conchologist* 26(2):5-6.

García, E. F. 1998. Some interesting Venezuelan molluscs - In 3 ft. of water or less. *American Conchologist* 26(4):8-9.

García, E. F. 1999. Observations on some New World mollusks. *American Conchologist* 27(2):4-6.
García, E. F. 1999. New molluscan records for the northwestern Gulf of Mexico. *American Conchologist* 27(2):27- 28.
García, E. F. 1999. Three new gastropod species from the New World. *Apex.* 14(3-4):59-65 (Stated date: December 20, 1999)
García, E. F. 2000. A confusing muricid species from the Western Atlantic. *American Conchologist* 28(2):14.
García, E. F 2000. Surprising new molluscan records from Louisiana and the northwestern Gulf of Mexico. *American Conchologist* 28(3):5-6.
García, E. F. 2000. Description of a very distinct *Cirsotrema* from New Caledonia *Novapex* 1(3-4):105-107 (Stated date: October 20, 2000).
García, E. F. 2000. A rare pair. *American Conchologist* 28(4):7.
García, E. F. 2001. A remarkable deep-water *Cochlespira* from Golfo de Chiriquí, southwestern Panamá. *The Festivus* 33(3):23-24.
García, E. F. 2001. A new *Nassarius* species from Golfo de Chiriquí, southwestern Panamá. *Novapex* 2(1): 25-27 (Stated date: March, 2001).
García, E. F. 2001. The saga of three ranellid species in the western Atlantic. *American Conchologist* 29(2):11-12.
García, E. F. 2001. Three new deep-water epitoniid species from the Philippines. *Novapex* 2(3):109-114 (Stated date: September, 2001).
García, E. F. 2001. A new species of *Latirus* from the southern Philippines. *Novapex* 2(4):149-151 (Stated date: December, 2001)
García, E. F. 2002. More discoveries from a collecting expedition off the Louisiana coast. *American Conchologist* 30(1):6-8.
García, E. F. 2002. Unexpected molluscan finds from the hydrocarbon vents off the Louisiana coast. *American Conchologist* 30(4):28-30.
García, E. F. 2003. New records of Indo-Pacific Epitoniidae (Mollusca: Gastropoda) with the description of nineteen new species. *Novapex* Hors Serie No.1, 1-22. (Stated date: February 10, 2003).
García, E. F. 2003. Three Argentine portraits. *American Conchologist* 31(1):12-13.
García, E. F. 2003. *Gyroscala turnerae*, a western Atlantic junior synonym of the Indo-Pacific *Gyroscala xenicima. American Conchologist* 32(2):21-23.
García, E. F. 2004. New records of *Opalia*-like mollusks (Gastropoda: Epitoniidae) from the Indo-Pacific, with the description of fourteen new species. *Novapex* 5(1):1-18.
García, E. F. 2004. On the genus *Cycloscala* Dall, 1889 (Gastropoda: Epitoniidae) in the Indo-Pacific, with comments on the type species, new records of known species, and the description of three new species. *Novapex* 5(2-3):57-68
García, E. F. 2005. Frustrations and extensions III. On *Columbella verrilli* Dall, 1889; and a geographic extension and generic placement of *Gymnobela malmii* Dall, 1881. *American Conchologist* 33(2):25- 27.
García, E. F. 2005. Six new deep-water molluscan species (Gastropoda: Epitoniidae, Conoidea) from the Gulf of Mexico. *Novapex* 6(4):79-87.
García, E. F. 2006. *Conus sauros*, a new *Conus* species (Gastropoda: Conidae) from the Gulf of Mexico. *Novapex* 7(2-3):71-76.
García, E. F. 2006. Six new species of mollusks (Gastropoda: Cerithioidea, Buccinoidea, Muricoidea) from Bahía de Campeche, southwestern Gulf of Mexico. *Novapex* 7(4):77-89.
García, E. F. 2007. Report on mollusks collected in a dredging expedition to Bahía de Campeche, southwestern Gulf of Mexico. American Conchologist 35(2):7764-11.
García, E. F. 2007. Results of deep-water dredging in the Gulf of Mexico using the "Benthic Skimmer", and report on several geographic extensions, including two species not previously reported in the western Atlantic. *The Festivus* 392:13- 18.

García, E. F. 2007. A new species of *Cosmioconcha* (Gastropoda: Columbellidae) from the northern Gulf of Mexico. *Novapex* 8(2):43-46.
García, E. F. 2007. Three new deep-water species of mollusks (Gastropoda: Calliostomatidae, Cystiscidae) from the southeastern Gulf of Mexico. *Novapex* 8(3-4): 27-133.
García, E. F. 2008. Eight new molluscan species (Gastropoda: Turridae) from the western Atlantic, with the description of two new genera. *Novapex* 9(1):1-15.
García, E. F. 2008. On two *Opalia*-like epitoniids (Gastropoda: Epitoniidae) from the Gulf of Mexico. *The Festivus* 40(7):83-86.
García, E. F. 2008. Frustrations and extensions IV: On two *Cyphoma* species from the Gulf of Mexico and six geographical extensions. *American Conchologist* 36(1): 4-8.
García, E. F. 2008. An extension of the genus *Spinosipella* (Bivalvia: Verticordidae) in the Gulf of Mexico. *American Conchologist* 36(3):8-9.
García, E. F. 2008. Four new buccinid species (Gastropoda: Buccinidae) from the western Atlantic. *Novapex* 9(4):141-148.
García, E. F. 2008. Molluscan findings from a recent dredging expedition off the Louisiana coast. *American Conchologist* 36(4):4-9.
García, E. F. 2009. A new *Astyris* species (Gastropoda: Columbellidae) from the Gulf of Mexico, with notes on the genus. *Novapex* 10(1):5-8.
García, E. F. 2009. Sorting out the generic placement of the three Panamic species assigned to *Eumetula* Thiele, 1912. *The Festivus* 41(5):43-46.
García, E. F. 2009. Book Review. Marine Shells of northeast Florida. *The Nautilus* 123(3):220-222.
García, E. F. 2010. *Costoanachis setularium* (d'Orbigny, 1839) in the Gulf of Mexico. *American Conchologist* 38(1):4-5.
García, E. F. 2010. The power of one. *American Conchologist* 38(1):6.
García, E. F. 2010. Bocas del Toro revisited. A follow-up of Olsson & McGinty's report on the Panamanian Archipelago. *American Conchologist* 38(2):4-12.
García, E. F. 2010. *Conus archetypus* Crosse, 1865 in northwestern Panama. *American Conchologist* 38(3):4-5.
García, E. F. 2010. A geographic extension for two species of *Favartia* (Muricidae: Muricopsinae) from the western Atlantic. *American Conchologist* 38(3):10-11.
García, E. F. 2010. The genus *Cirsotrema* (Gastropoda: Epitoniidae) in the Panamic Province, with the description of two new species. *Novapex* 11(4):93-98.
García, E. F. 2010. Description of four new species of *Cochlespira* (Gastropoda: Turridae) from the New World. *Novapex* 11(4):107-113.
García, E. F. 2011. A new species of *Eccliseogyra* (Gasropoda: Nystiellidae) from southeastern Brazil. *The Nautilus* 125(3)167-170.
García, E. F. 2011.Sea star stomach uncovers geographical extension for *Parvanachis guerreroensis* Strong & Hertlein, 1937, and other goodies. *The Festivus* 43(9):93-94.
García, E. F. 2011. A new species of *Mitra* (*Fusimitra*) (Gastropoda: Mitridae) from the northwestern Gulf of Mexico. *Novapex* 12(3-4):57-62.
García, E. F. 2011.Two new species of *Epitonium* (Gastropoda: Epitoniidae) from the western Atlantic. *Novapex* 12(3-4):99-107.
García, E. F. 2011. Noteworthy mollusks of the Gulf of Mexico, including record sizes and geographical extensions. *American Conchologist* 39(3):4-9.
García, E. F. 2011. On *Vexillum (Pusia) articulatum* (Reeve, 1845) and *V. (P.) trophonium* (Dall, 1889). *American Conchologist* 39(3):10-11.
García, E. F. 2012. Preliminary report on the possible effect of the Deepwater Horizon oil spill in the Gulf of Mexico on the molluscan fauna of the surrounding area. *American Conchologist* 40(1):32-33.

García, E. F. 2012. Noteworthy offshore mollusks from the north- central Gulf of Mexico, including geographical extensions and a generic reassignment. *American Conchologist* 40 (1): 34- 37.

García, E. F 2012. *Terebra limatula* Dall, 1889 and *T. acrior* Dall, 1889 (Gastropoda: Terebridae); two problematic taxa from the western Atlantic. *Zootaxa* 3328:66-68.

García, E. F 2012. Report on a dredging expedition off the Louisiana coast including geographical extension and new record size. *American Conchologist* 40(4):7-10.

García, E. F 2013. Report on geographical extensions and record-size specimens of *Fenimorea*, *Leptodrillia* and *Cerodrillia* (Conoidea: Drilliidae) from the Gulf of Mexico. *The Festivus* 45(1):3-5.

García, E. F 2013. Book Review. Monograph of the little slit shells. *The Festivus* 45(2):15-16.

García, E. F 2013. Report on a dredging expedition off the Louisiana coast including geographical extension and new record size. II. *American Conchologist* 41(1):4-8.

García, E. F 2013. On the generic placement and identity of *Timbellus phaneus* (Dall, 1889) and *T. havanensis* (E. H. Vokes, 1970) (Gastropoda: Murididae). *American Conchologist* 41(4):4-7.

García, E. F 2013. *Turritella lyonsi* found off the Louisiana coast. *American Conchologist* 41(4):8-9.

García, E. F 2015. A new species of *Zafrona* (Gastropoda: Columbellidae) from the Gulf of Mexico. *Novapex* 16(2):49-52.

García, E. F 2015. A new species of *Latiromitra* (Gastropoda: Ptychatractidae) from the Gulf of Mexico. *The Nautilus* 129(2):90-93.

García, E. F 2015. A new species of *Glyphostoma* (Gastropoda: Clathurellidae) from the Gulf of Mexico. *The Nautilus* 129(3):136-139.

García, E. F 2016. A new species of *Epitonium (LIbrariscala)*(Gastropoda: Epitoniidae) from French Polynesia. *Novapex* 17(1):5-8.

García, E. F 2016. The genera *Miraclathurella* Woodring., 1928(Gastropoda: Pseudomelatomidae) and *Darrylia* García, 2008 (Gastropoda: ?Horaiclavidae), with two proposed new combinations for Darrylia. *The Nautilus* 130(2):79-81.

García, E. F 2016.The western Caribbean complex of *Fasciolaria tephrina* de Souza, 2002 (Gastropoda: Fasciolariidae), with the description of a new species. *The Nautilus* 130(3):127-131.

García, E. F 2018. Report on a dredging expedition off the Louisiana coast including surprising geographical extensions. *American Conchologist* 46(3):8-12.

García, E. F 2019. Report on a dredging expedition off the Louisiana coast including surprising geographical extensions II. *American Conchologist* 47(4):4-11.

García, E. F 2020.*Cyphoma aureocinctum* (Dall, 1889), a remarkable find on a Louisiana beach. *American Conchologist* 48(1):30-31.

García, E. F 2021. Report on mollusks collected in a single dredge haul in the Gulf of Mexico. What does it say about future molluscan discoveries? *American Conchologist* 49(2):16-21.

García, E. F 2021. Species of the family Borsoniidae dredged by the R/V *Pelican* in the Gulf of Mexico. *American Conchologist* 49(4):8-12.

García, E. F 2021. A new species of *Bathytoma* (Gastropoda: Conoidea: Borsoniidae) from the western Caribbean Sea. *The Nautilus* 135(3-4):89-91.

García, E. F 2022. On two species of *Amaea* (Gastropoda: Epitoniidae) from the western Atlantic Ocean. *American Conchologist* 50(2):8-11.

García, E. F 2022. Species of Buccinidae collected by the R/V *Pelican* in the Gulf of Mexico. *American Conchologist* 50(1):20-23.

García, E. F 2022. Species in the families Cystiscidae, Marginellidae, and Granulinidae dredged by the R/V *Pelican* in the Gulf of Mexico. *American Conchologist* 50(4):20.

García, E. F 2023. Species of the families Costellariidae and Mitridae dredged in the Gulf of Mexico by the R/V *Pelican* in the Gulf of Mexico. The Festivus 55(1):65-70.

García, E. F 2023. Species of Muricidae (Gastropoda: Muricoidea) dredged by the R/V *Pelican,* and the synonymy of two well- known names. *Novapex* 24(1):21-32.

García, E. F 2023. Species of the families Conidae and Terebridae dredged by the R/V *Pelican* in the Gulf of Mexico. *American Conchologist* 51(1):11-22.

García, E. F 2023. *Falsuszafrona yucatecana* (Gastropoda: Columbellidae), a new species from the southwestern Gulf of Mexico. *The Nautilus* 137(2):52-54.

García, E. F 2023. Species of the family Fissurellidae dredged by the R/V *Pelican* in the Gulf of Mexico. *American Conchologist* 51(3):4-8.

García, E. F. 2023. Species of Fasciolariidae, Dolicholatiridae, Busyconidae and Coumbellidae dredged by the R/V *Pelican* in the Gulf of Mexico. *Gloria Maris* 62(2):58-66.

García, E. F. 2023. A new species of *Fenimorea* (Gastropoda: Drilliidae) from the eastern Gulf of Mexico. *The Festicus* 55(4):246-248.

García, E. F. 2023. Species of the superfamilies (Gastropoda: Areneidae, Calliostomatidae, Colloniidae, Liotiidae, Skeneidae, Solariellidae, Trochidae, Turbinidae) and Seguenzioidea (Gastropoda: Calliotropidae, Cataegidae,Chilodontidae, Cataegidae, Chilodontidae, Seguenzidae) dredged by the R/V *Pelican* in the Gulf of Mexico. *Xenophora Taxonomy* 41(17- 23).

García, E.F. A new species of *Fenimorea* (Gastropoda: Drilliidae) from the eastern Gulf of Mexico. The Festivus 55(4):238.

García, E. F. 2024. A new species of *Corinnaeturris* (Gasropoda: Clathurellidae) from the southeastern Gulf of Mexico. The Festivus 56(1):27-30.

García, E. F. 2024. Species of the families *Epitoniidaei and* Architectonicidae *(Mollusca: Gastropoda) dredged* by the R/V *Pelican* in the Gulf of Mexico. *Novapex* 25(1):1-9.

García, E. F. 2024. A shell collector in a nut shell. American Conchologist 52(1):29.

García, E. F. 2024. Species of the families Drilliidae Olsson, 1964 and Clathurellidae H. Adams & A. Adams, 1858 (Gastropoda: Conoidea), collected by the R/V *Pelican* in the Gulf of Mexico. *American Conchologist* 52(3):10-16.

García, E. F. and W. K. Emerson. 1987. A new species of *Scaphella* (Gastropoda: Volutidae) from off Yucatán, Mexico. *Apex. Informations Scientifiques de la Société Belge de Malacologie*, 11(1):1-5.

García, E. F. and K. Sunderland. 1991. Notes on the genus *Lyria* in the western Atlantic. *American Conchologist* 18(l):18-19.

García, E. F. and H. G. Lee. 2002. Report on molluscan species found in the offshore waters of Louisiana, including many extensions of known range and un-named species. *American Conchologist* 30(4):10-13.

García, E. F. and H. G. Lee. 2003. Report on molluscan species found in the offshore waters of Louisiana, including many extensions of known range and un-named species. II. *American Conchologist* 31(1):26-29.

García, E. F. and H. G. Lee. 2004. Report on the malacofauna of offshore Louisiana waters - including many range extensions and un-named species. III. *American Conchologist* 32(3)::21-24.

Garcia, E. F. and E. H. Vokes. 2014. A mystery solved. *American Conchologist* 42(4):22-23.

García, E. F., W. G. Lyons and M. A. Snyder. 2016. The western Caribbean complex of *Fasciolaria tephrina* de Souza, 2002 (Gastropoda: Fasciolariidae), with the description of a new species. *The Nautilus* 130(3):127-131.

García, E.F. and R. Salisbury, 2023. *Panamitra* (Gastropoda: Mtridae), a new genus from the Panamic Province with a unique radula in the family Mitridae. *Conchylia* 54(1-2): 15-23.

García, E. F. and R. Salisbury. 2024. Additions to the genus *Panamitra*. *Conchylia* 23(3-4):20.

Rosenberg, G., F. Moretzsohn & E. García. 2009. Gastropoda (Mollusca) of the Gulf of Mexico. *In*: *Gulf of Mexico: Its Origins, Waters, and Biota.II. Biodiversity.* D. L. Felder & D. K. Camp, eds, Texas A & M University Press, pp. 579-699.

Salisbury, R, E. F. García & A, Fedosov. 2023. On *Mitra sulcata* Swanson in G. B. Sowerby I, 1825 (Gastropoda: Mitridae), type species of *Subcancilla* Olsson & Harbison, 1952: re-description, selection of neotype, and radular information. *Conchiglia* 53 (3-4).

PLAYS DIRECTED:

"The House of Bernarda Alba," Federico García Lorca, Lafayette Little Theater, December, 1968.
"Los Fantoches," University of Southwestern Louisiana Student Union, November, 1974.
"La Guarda Cuidadosa," Miguel de Cervantes, University of Southwestern Louisiana Student Union, April, 1976.
"Esta noche juntos, amándonos tanto," Maruxa Vilalta, University of Southwestern Louisiana Student Union, March, 1978.
"Falso Drama," Rodolfo Usigli, University of Southwestern Louisiana Student Union, March, 1979.

OTHER ACTIVITIES:

Past President, Acadiana Bromeliad Society.
Past President and founder, Louisiana Malacological Society.
Speaker on many occasions to area civic groups and schools on bromeliads, Latin America and molluscs.
Speaker on many occasions at the International Convention of Conchologist of America.
Choreographer of Spanish-American folk dances for several years.
Scientific judge at shell shows in St. Petersburg, Florida, Sarasota, Florida; Indianapolis, Indiana, St. Louis, Missouri; Sanibel Island, Florida; Melbourne, Florida and Jacksonville, Florida; Wilmington, North Carolina.
Designer of the definitive issue of stamps on seashells for the country of Belize.

- https://www.conchology.be/
  https://www.conchology.be/?t=9001&id=18957&srsltid=AfmBOoqWSknnAFzq_8fKgZJYv2WBV4P4hzKU0qGNOdbO0de6RzYl1u_F
- https: jaxshells.org

www.ingramcontent.com/pod-product-compliance
Lightning Source LLC
LaVergne TN
LVHW071241210926
848486LV00007B/316

* 9 7 9 8 8 9 7 6 6 9 4 8 6 *